Maurício Santos Araújo

Challenges and possibilities in biology teaching

Maurício Santos Araújo

Challenges and possibilities in biology teaching

Teacher-student interaction in the classroom

ScienciaScripts

Imprint

Cover image: www.ingimage.com

This book is a translation from the original published under ISBN 978-613-9-66397-2.

Publisher:
Sciencia Scripts
is a trademark of
Dodo Books Indian Ocean Ltd. and OmniScriptum S.R.L publishing group

120 High Road, East Finchley, London, N2 9ED, United Kingdom
Str. Armeneasca 28/1, office 1, Chisinau MD-2012, Republic of Moldova, Europe
Printed at: see last page
ISBN: 978-620-8-02776-6

SUMMARY

CHAPTER 1

CONTRIBUTIONS OF THE INSTITUTIONAL TEACHING INITIATION SCHOLARSHIP PROGRAMME (PIBID) TO THE TRAINING OF BIOLOGICAL SCIENCES UNDERGRADUATES AT THE FEDERAL INSTITUTE OF PIAUÍ

MAURÍCIO DOS SANTOS ARAÚJO [mauriciosangesl 1@hotmail.com]

Master's student in Genetics and Improvement at the Federal University of Piauí. He is currently working on Quantitative Genetics (genetic improvement of cowpea) and Toxicological Genetics with toxic, cytotoxic, genotoxic and mutagenic evaluations of medicinal plants. He develops projects aimed at teaching Genetics, with an emphasis on the construction of didactic-pedagogical materials.

Summary

The Institutional Teaching Initiation Scholarship Programme (PIBID) seeks to promote the appreciation of the teaching profession by providing scholarship holders with a space to build teaching skills and abilities. With this in mind, the aim of this article is to analyse the contributions made by the Biology sub-project, developed under the PIBID programme, to the teaching training of scholarship holders at the IFPI, Floriano Campus. The theoretical and methodological bases of this study are based on a qualitative and quantitative perspective, with a descriptive approach. The collection instrument used was a form with open and closed questions, linked to the *Google Forms* platform, with 17 (seventeen) PIBID/IFPI/ Biology subproject scholarship holders from the Floriano *campus*. The results show that, after joining the programme, the students increased their performance on the course, especially in terms of behaviour and posture during seminars, which many pointed out as a factor that hindered the acquisition of more knowledge. They believe that the programme values teacher training and, due to the nature of its contribution, suggest expanding it to a larger number of undergraduates. At the end of the research, it was found that the programme is of fundamental importance for the training of undergraduates, as it provides an ideal space for building knowledge in the educational environment.

Keywords: PIBID; Teacher training; Valorisation of the teaching profession.

INTRODUCTION

Over the years, education professionals have sought better working conditions in Brazilian schools from the government (Leal & Cardoso, 2015). However, several obstacles are encountered during this process. Professional demotivation is one of them, as it has become one of the factors most pointed out by education professionals, who list: overcrowding in classrooms, overload of activities, lack of professional recognition, work environment without adequate infrastructure, job insecurity, low financial incentives, among other factors that can hinder the teacher's pedagogical practice. In addition, these aspects can contribute to the failure of students to seek degree programmes as a mechanism for choosing their future profession (Giacomelli, Borges & Santos, 2016).

For teachers to be able to innovate in the classroom, they need to constantly motivate themselves. But for this to happen, the school must provide the minimum conditions necessary for teachers to carry out dynamic, investigative and innovative work, using active methodologies and technological resources in their practice (Araújo, Sousa & Cruz, 2016). This is why preparation is essential, because *"what does a classroom need for quality education? It fundamentally needs well-prepared, motivated and well-paid teachers with up-to-date pedagogical training. This is indisputable"* (Moran, 2004, p.15). In this sense, aspects such as planning, teaching resources and motivation can contribute to innovation in the classroom, providing students with something different during the lesson.

In view of the devaluation of teaching professionals, various educational policies have been created to encourage academics to enter and remain in teaching careers, one of them being the Institutional Teaching Initiation Scholarship Programme (PIBID). The programme is part of a

a national policy to enhance the initial training of teachers to work in basic education and thus raise the quality of Brazilian public education. It was created through the Ministry of Education's (MEC) Normative Ordinance No. 38 on 12 December 2007, with the aim of encouraging teacher training and contributing to valuing the teaching profession (Brasil, 2010; Frison, Veiga-Simão & Cigales, 2017). In this way, the programme's aim is to bring undergraduates into the school environment, create a link between basic education and Higher Education Institutions (HEIs), in order to contribute to the articulation of knowledge between theory and practice (Brasil, 2012; Ribeiro, 2013).

PIBID provides those involved, specifically the scholarship holders, with a dialogue between the various areas of knowledge, thus promoting direct contact with the future *locus of* action (Hemielewski & Canton, 2014). In this way, the exchange of knowledge that makes up professional training is based on objectives that are described in Normative Ordinance No. 38, of 12 December 2007, which, among various aspects, aims to:

> *"I - to encourage the training of teachers at higher education level for basic education; II - to contribute to the valorisation of the teaching profession; III - to raise the quality of initial teacher training in degree courses, promoting integration between higher education and basic education; IV - to include undergraduates in the daily life of schools in the public education network, providing them with opportunities to create and participate in methodological, technological and teaching practice experiences of an innovative and interdisciplinary nature that seek to overcome problems identified in the teaching-learning process; V - encourage public basic education schools, mobilising their teachers as co-trainers of future teachers and making them protagonists in the processes of initial training for teaching; VI - contribute to the articulation between theory and practice necessary for teacher training, raising the quality of academic actions in undergraduate courses; VII - contribute to undergraduate students becoming part of the school culture of teaching, through appropriation and reflection on instruments, knowledge and peculiarities of teaching work" (Brasil, 2013, p. 2-3). 2-3).*

PIBID is subordinate to the Coordination for the Improvement of Higher Education Personnel (CAPES), one of whose aims is to promote teacher training. It also seeks to support the realisation of interdisciplinary projects in basic education, promoting a link between higher education institutions and basic education (Brasil, 2012). In this sense, the National Education Guidelines and Bases Law (LDBEN) No. 9.394, of 20 December 1996, in its article 62, discusses teacher training, specifically in paragraph 5[2] , states that it is the duty of all federal entities to encourage the training of teaching professionals to work in basic education (Brasil, 2015).

The programme allows scholarship holders to have direct contact with their future workplace through practical activities that are carried out in partner schools, as the teacher in training can improve their teaching methodologies, conceptions and objectives that are linked to the teaching career (SALLAI, 2016). For this reason, the programme, through interdisciplinary activities, allows for dialogue with other areas of knowledge, as it is an educational policy that values teacher training so that they can awaken a reflective stance on what it means to be a teacher (Dorneles, Galiazzi & Altenhofe, 2017). In this sense, Pimenta (2002, p.7) discusses the new identity of the teacher, with regard to being and doing teaching:

> *"A teacher's professional identity is built on the social meaning of the profession [...] it is also built on the meaning that each teacher, as an actor and author, gives to the teaching activity of situating themselves in the world, from their life history, their representations, their knowledge, their anxieties and desires, the meaning they have in their life: being a teacher. As well as from their network of relationships with other teachers, in schools, unions and other groupings".*

The programme is divided into sub-projects that cover various areas of knowledge, such as: Arts, Biology, Physics, History, Chemistry, Mathematics, Music, Physical Education and others (Ramos, Araújo & Santana, 2016). With regard to the Biology subproject, activities are developed that facilitate understanding of the biological processes that underpin life. Therefore, there is a need to create didactic-pedagogical materials that lead students to relate certain content, previously considered merely theoretical, thus contributing to a broader understanding of the biological concepts that permeate students' lives (Viana et aL, 2016).

Many authors attribute the success of PIBID as a programme to enhance teacher training to the granting of scholarships to undergraduate students, supervisors (basic education) and coordinators (university). This systemic relationship aims to find ways to improve Brazilian education, as well as making it possible for basic education teachers to return to university in order to update their professional skills (Chaluh et al., 2017). In this way, the supervisor, together with the partner school, where the undergraduates develop the interdisciplinary projects, are responsible for the training of the scholarship holders, as *they "(...) accompany the activities of the scholarship holders in the school space, thus acting as co-trainers in the process of initiation to teaching, in conjunction with the university trainer.* " (Gatti et al, 2014, p. 10).

According to the Law of Guidelines and Bases of National Education (LDB) No. 9.394, of 20 December 1996, in Title VI on Education Professionals, it discusses in art. 62, § 5º that all federative entities should encourage the training of teachers in basic education, through the PIBID programme for undergraduate students at higher education institutions (Brazil, 2016). From this perspective, teacher training starts from two guiding points: theoretical-scientific training and technical-practical training, as these aspects aim to prepare the education professional to develop their didactic-methodological skills, improve their training practice, investigate educational and epistemological aspects, among others (Libâneo, 1994).

The quality of public education will only be achieved when there is real investment in improving future teachers, and for this to happen they need to be introduced to the *workplace* very early on. For this reason,

"[...] *in the ongoing training of teachers, the fundamental moment is that of critical reflection on practice. It is by thinking critically about today's or yesterday's practice that the next practice can be improved"* (Freire, 2011, p.40). When teachers are able to build reflective thinking about their pedagogical importance in the student's civic education, they will understand the genesis of their mistakes and devise strategies to correct them in order to make the education process increasingly egalitarian and integrated. Therefore, when professionals are able to reflect on their practice, this attitude *"ennobles the teaching profession"* (Tardif, Borges & Maio, 2012, p. 11).

Teacher training is a subject discussed by many scholars in the field of education. Thus, authors such as Dewey (1958), Nóvoa (1991), Libâneo (1994), Schõn (2000), Tardif (2002), Demo (2004), Arroyo (2005), Zabala (2008), Saviani (2009), Carvalho & Gil-Pérez (2011), Freire (2011), Pimenta (2012) and others have engaged in dialogue on this subject. In this sense, teachers need to build a solid theoretical base during their training, putting theory into practice. Only in this way will they be able to acquire new knowledge in order to work in their profession in a critical and reflective way. For this reason, Freire (1998, p. 44) states that:

> *"[...] in ongoing teacher training, the key moment is critical reflection on practice. It is by critically thinking about today's or yesterday's practice that the next practice can be improved. The theoretical discourse itself, which is necessary for critical reflection, must be so concrete that it almost blends in with practice."*

When teachers take a critical stance in their pedagogical practice, without reproducing what the educational system often offers, they can, through critical reflection, turn the classroom into an environment of transformation. Therefore, when teachers re-evaluate their practice, seeking to identify the positive and negative factors that emerge from it, only then will they be able to concretise the formation of educational precepts that are linked to ethical and professional paradigms (Barreiro & Gebran, 2006). Thus, when it comes to Biology as an area of knowledge, there is a need to introduce new teaching methodologies, so that there is a gap between the purely expository classes left by the traditional teaching current, which, many times, this type of class may not corroborate with the effective association between theory and practice (Cardoso, Araújo & Sousa, 2016; Lima et al., 2016). Therefore, carrying out projects that aim to relate theoretical foundations to practice can lead students to more effective learning, given that Biology is an area of knowledge with a wide range of biological processes, scientific terminology and merely practice (Viana et al., 2016).

The realisation of interdisciplinary projects contributes to the construction of knowledge linked to the area of Biology, which is why Oliveira (2006, p. 16) corroborates that:

> "*Working with projects shifts the focus of the classroom from the teacher to the student, from information to knowledge, from memorisation to learning. It balances theory and practice, divides responsibilities and tasks, communicates results and discusses assessment processes. When working with projects, teacher and student take on the role of researchers and co-responsible for the learning process.* "

With this in mind, this article aims to analyse the contributions of the Biology subproject of the Federal Institute of Education, Science and Technology of Piauí - IFPI, Floriano *Campus*, to the teacher training of the scholarship holders who are members of the Biology subproject.

METHODOLOGY

This research was carried out with Biological Sciences undergraduates taking part in the Institutional Teaching Initiation Scholarship Programme (PIBID), Biology subproject, linked to the Federal Institute of Education, Science and Technology of Piauí - IFPI. The sample group was 17 (seventeen) PIBID scholars working in 02 (two) state schools, which are partners in the programme: the Osvaldo da Costa e Silva Normal School (ENOCS) and the Fauzer Bucar School Unit, located near Praça Sobral Neto, S/N - Centro, CEP: 64800-000 in Floriano/PI. It is worth noting that, in order to preserve the identity of the subjects participating in this investigation, fictitious names were used.

The theoretical and methodological bases of this study are based on a qualitative and quantitative perspective, with a descriptive approach. Data was collected using a form with open and closed questions that sought to identify the contributions made by PIBID to teachers in training. This instrument was made available through the electronic platform *(Google forms)*. The use of this tool in education research is of great value, as it provides agility and practicality when collecting and tabulating data. As well as filtering all the responses, it can also generate graphs to illustrate the results (Heidemann & Oliveira, 2010).

We used the content analysis proposed by Bardin (2011), which used criteria such as: i) history and theory (historical perspective); ii) practical part (analysis of mass communication, open questions and tests); iii) methods of analysis (organisation, coding, categorisation, inference and computerisation of analysis) and iv) analysis techniques (categorical analysis, evaluation, enunciation, propositional discourse, expression and relationships). In this sense, the procedure for analysing and discussing the qualitative data was based on the descriptive interpretation of the fellows' statements, preserving the essence of their speeches. The *Microsoft Office Excel 2016®* programme was used to construct the graphs and tables, as this software seeks to create instruments that facilitate understanding of the data collected in the research (Ribeiro Júnior, 2013).

Legal aspects of research

The research was supported by current Resolution 510 of 07 April 2016, based on the regulations and legal attributions of Law No. 8.080 of 19 September 1990, Law No. 8.142 of 28 December 1990 and Decree No. 5.839 of 11 July 2006, which deals with ethics in research with human beings. During the methodological process, the participants were presented with the justification, objectives, methodology and possible benefits, and at no time did the research present any risk to the participants. Their ideology, dignity and autonomy were respected in order to preserve their privacy, as well as guaranteeing the confidentiality of all personal information (Brasil, 2016).

RESULTS AND DISCUSSIONS

The PIBID programme aims to involve undergraduates in the school environment from an early age, through interdisciplinary projects. In this way, there is an exchange of knowledge between the scholars and the students of the partner school, as both benefit from the pedagogical process. In this sense, we sought to find out how the PIBID scholarship holders performed before joining the programme, in terms of staying on the course, interacting with other students, performing well in written tests, holding seminars, writing projects and scientific articles and relating to the public in terms of orality.

> *"Before I joined the PIBID programme, my performance in the classroom was very good, but I found it difficult to argue and present my ideas on a given subject in public, because I didn't put this into practice routinely. " (Scholar Mauro).*
>
> *"(...) PIBID helped me a lot to stay on the course, because it's a programme that brings together people from various modules and whenever we did projects together I always learned something more. I was able to learn many things from the fellows on the following modules, in relation to subjects I hadn't seen, PIBID helped me a lot, because I built up prior knowledge about certain Biology content. "* (Student Very).
>
> *"I used to find it difficult to contextualise the content, unlike today when I produce projects aimed at secondary school students, which has given me many meaningful experiences. "* (Andressa).
>
> *"I was a bit undefined, because I hadn't had any contact with teaching properly, direct contact with the classroom, I didn't know how to plan and operationalise a project, because I was very lay. So I didn't develop in the course as I should have before PI BI D." (Maria).*
>
> *"I had little methodological vision for working with biology in the classroom and I had no affinity with the degree. "* (Scholar Ivo).

The PIBID makes it possible to articulate knowledge, as it gives the scholarship holders interaction with their future place of work. This is only possible thanks to the realisation of interdisciplinary projects, allowing the scholarship holders to develop their teaching skills and abilities, mainly by helping them during their initial training (Maldener, 2007). In addition, the programme seeks to build a broad understanding of being a teacher, helping them to understand the real reason for teaching. In this way, the PIBID programme plays an important role in training, because from the moment the student is inserted into the educational context, they assimilate the difficulties that emerge in the teaching-learning process, but can, through reflective thinking, seek possible solutions to minimise these problems, contributing to the construction of their professional identity (Massena & Siqueira, 2016).

Students who take part in the PIBID programme have a range of possibilities in terms of preparing for a teaching career, as the programme not only makes it possible to carry out interdisciplinary projects, but also encourages scholarship holders to develop scientific writing through reports, projects and scientific articles, in order to foster a broader education. Therefore, the PIBID scholarship holders were asked how they were performing in their Biology subjects after joining the programme and what changes they and others had noticed after their time in the programme.

> *Today I find it easier to speak in public because I already have direct contact with teachers and students. "* (Scholar Mariana).

"The teachers on the Biology course said that we had improved a lot in terms of posture and development in seminar presentations. One of the difficulties I had was presenting seminars in class and I'm overcoming it with PIBID. " (Iara Cristine).

"Direct contact with the programme's partner school provides fundamental experiences for teaching, opening up a range of knowledge during and after our academic life. Therefore, helping to develop practices as a future teacher. " (Scholar Leandro/

"*PIBID provides an environment for the scholarship holders to develop interdisciplinary projects as a group and this helps a lot with their academic performance. "* (Scholar Barjonas).

"(...) it was through PIBID that I had direct contact with a classroom for the first time, I was able to see the reality of school, and then when I started my supervised internship I already had experience with students and I was able to improve my pedagogical practice (...)." (Maria).

"It's extremely important, because as a future educator I already know a little about the reality of public schools and I already know how to develop projects in order to improve my students' learning through interaction between teacher and student. " (Júlio).

"With PIBID we can better experience the routine of a school and already learn how to deal with problems that may arise and deal directly with students and their different ways of behaving and learning." (Juliano).

"It provides a rich experience, bringing a new perspective on the profession of being a teacher. " (Fernanda).

The contributions of the PIBID programme are currently being discussed on a national level. The inclusion of undergraduates in the school environment prior to their curricular internships contributes to the development of teaching methodologies. In addition, the PIBID programme encourages continuing teacher training, creating a link between basic education schools and higher education institutions, leading to projects that facilitate the interrelationship between educational institutions (Sallai, 2016). In this sense, it is necessary that during teacher training there is a union between theoretical and practical elements, so that the teacher in training can put into practice all the theoretical apparatus built up at university (Libâneo & Pimenta, 1999).

The aim was to identify how the link between theory and practice is being built, based on the interdisciplinary projects carried out within the scope of the Biology subproject at IFPI, Campus-Floriano. Among the students surveyed, 18 per cent said that this link was being made on a regular basis; 53 per cent said that this correlation was being made in an excellent way and 29 per cent of the students considered this link between theory and practice to be excellent, as shown in figure 1.

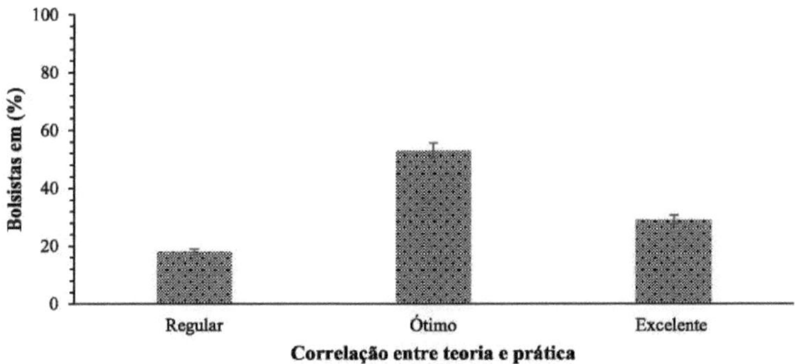

Figura 1 - Articulation between theory and practice according to the scholarship holders' perspective.

The construction of learning presupposes the student's experiences throughout the training process. For this reason, actions designed to help undergraduates develop their professionalising skills in terms of teaching and the school curriculum contribute to their qualification in basic education (Garcia, 1999). In this way, PI BI D has helped undergraduates to build new experiences linked to teaching, even before the start of their compulsory curricular internship, helping to build methodological skills in order to achieve the desired level of teaching practice (Felício, 2014).

With regard to the association between theory and practice, teachers must provide students with the right environment to develop their skills (Paranhos, Souza Filho & Paranhos, 2016). In this sense, it is necessary to awaken the student's concern in the face of the unknown, so that they can seek their own answers through problematisation, following logical and guiding principles in order to test them. In this way, they can build integrative learning, based on logical and theoretical principles (Bizzo, 2009). However, for this association to be made, it is necessary to break through the weaknesses found in the training of many teachers. For this reason, it is necessary to consider the inclusion of educational policies that provide undergraduates with the means to put this knowledge built up at university into practice (Machado, Vasconcelos & Oliveira, 2017).Regarding the experiences built up during the realisation of the projects, within the scope of the Biology subproject, we sought to identify what these contributions were from the students' point of view: 12% reported that these experiences are being built up on a regular basis. On the other hand, 70% consider the building of knowledge within the PIBID programme to be excellent; 12% said that these experiences are being built in an excellent way and 6% did not know how to answer, as shown in figure 2.

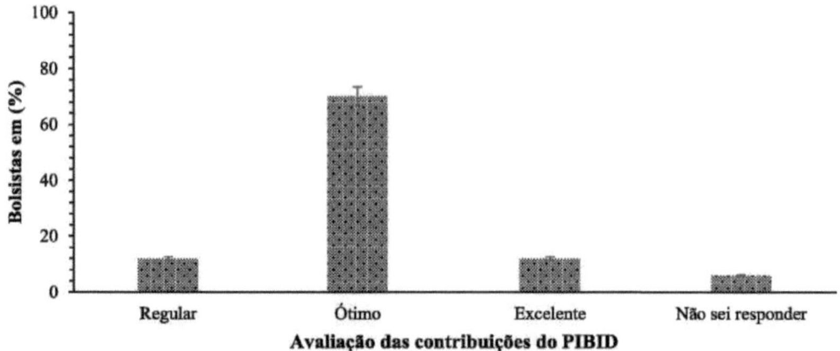

Figura 2 - Construction of experiences through the realisation of interdisciplinary projects developed within the scope of PIBID, according to the scholarship holders' vision.

One of the main problems currently faced in teacher training is the lack of practice and experience in everyday school life, which can contribute to poor performance during the teacher's pedagogical practice. Newly qualified teachers can have difficulties in their pedagogical practice because they often don't have effective classroom experience, a factor that can hinder their practice (Géglio & Silva, 2015). For this reason, the PIBID programme aims to encourage undergraduates to understand everyday life and learn how to resolve situations in the classroom, developing innovative and interdisciplinary projects so that, during their undergraduate studies, they can build new learning experiences with regard to being a teacher. From this perspective, the programme can provide the scholarship holder with this practical and methodological apparatus, leading them to a constructivist education (Abreu, 2012).

The PIBID programme is a public policy for valuing teacher training. With this in mind, we sought to identify the importance of the programme, together with the scholarship holders from the *\FP\-campus* Floriano Biology subproject, in terms of the improvements it provided them with during the course and the experiences built up for their training as future teachers.

> "The *programme is extremely important for my teacher training, it was through PIBID that I was able to have contact with a classroom for the first time. I was able to see the reality of school, and then, when I started my supervised internship, I already had experience with students and with some pedagogical practices that I was able to take to my internships. " (Scholarship holder João Paulo).*
>
> *"It's of the utmost importance for graduates, only then will we have quality public education. This way, we'll have teachers who are better prepared from the moment they graduate. " (Scholar Maria).*
>
> *"It's extremely important, because as a future educator I already know a little about the reality of public schools and I already know how to develop projects that make student-teacher interaction happen. All of this was only possible thanks to PIBID, because it provides this training based on the dialogue of knowledge. " (Maria Valentina).*

"With PIBID, we can better experience the routine of a school and learn from an early age how to deal with problems that arise and deal directly with students and their different ways of behaving and learning. " (Scholarship holder Francisco).

It has a strong influence, because it is during the internship that we have our first contact with the classroom, and in a schematic way. However, the PIBID allows us to be better prepared and develop professionally from the time we graduate. " (Fellow Iara Cristine)

PIBID has been of great importance in my academic life, having provided significant learning, as well as successful experiences in my professional training. " (Scholarship holder Andressa).

"My vision of what it would be like to teach a class has changed completely, I can say that I see the work of the teacher with different eyes and as a teacher who cares about how he is passing on content to the student, seeing if he is learning. I've built up a reflective mindset about being a teacher. " (Scholar Ivo).

In a similar study carried out with PIBID scholarship holders from the Biology subproject at IFPI in the city of Floriano-PI, it was found that the students, after

When they joined the programme, they improved their performance in many areas. In this sense, 46.7% of the students corroborated that PI BI D contributed to a significant improvement in their performance, due to the direct contact with the classroom and the experiences gained in the projects that are developed in the partner schools. For this reason, they consider the programme to be an excellent public policy aimed at training teachers to work in basic education (ARAÚJO et al., 2015).With regard to teacher training, PIBID works in conjunction with higher education institutions with a training mission. In this way, the scholarship holders were asked to list the following factors from 1 to 10, according to their degree of importance: direct contact with their future place of work, their improvement in higher education in terms of performance in subjects, the importance of carrying out interdisciplinary projects, the PIBID as an instrument for training teachers for basic education, the production of scientific work, financial aid for scholarship holders, compliance with the action plan governing the projects and the construction of knowledge based on the projects carried out, as illustrated in Table 1.

Table 1 - Positive points of the PIBID programme in terms of teacher training.

Issues discussed	**Level of importance (1-10) of the positive points of the PIBID Programme**									
	1	**2**	**3**	**4**	**5**	**6**	**7**	**8**	**9**	**10**
Direct contact with the classroom.	*	*	*	*	*	*	47%	12%		41%
M elhoriano course performance.	*	*	*	*	12%	5%	5%	24%	30%	24%
Interdisciplinary projects.	*	*	*	*	*	*	29%	24%	12%	35%
Teacher training.	*	*	*	*	*	6%	12%	29%	29%	24%
Production of articles, projects and reports.	*	*	*	*	6%	*	12%	29%	29%	24%
Financial support.	*	*	*	*	6%	6%	6%	6%	12%	64%
Compliance with the action plan.	*	*	*	*	*	6%	6%	18%	29%	41%

Building knowledge.	*	*	*	*	*	*	*	18%	12%	70%

* Graphical representation assigned to the table in the spaces where the scholars did not assign any value to the assertion, assigning the percentage equal to zero.

The PIBID programme gives students a grant of R$400.00 (four hundred reais) a month for a period of 2 (two) years, with the possibility of renewing it for the same period. This aid is intended to enable them to travel to the partner school where the subproject operates. The projects carried out in the schools are developed by adapting the subproject in the action plan, which contains all the activities that will be carried out by the scholarship holders during a school year. The entire practical implementation of the projects is under the guidance of the area supervisors, who are subordinate to the area coordinator, in order to seek out strategies with the partner school that facilitate student learning and, in doing so, give the scholarship holders experience as a mechanism to encourage teacher training (Brasil, 2010).

In an effort to find mechanisms to improve the PIBID programme, the fellows were asked what suggestions they could make in order to find mechanisms to improve PIBID's work in the partner schools, as well as suggestions to improve the programme, both institutionally and personally.

> *"My suggestion is that everyone at the partner school should also be involved in the projects carried out by the fellows and supervisors so that better results can be achieved for all the students. And that PIBID should not just be a programme for a few students and without support from the school management, but that it should arouse the interest of other professionals at the school.* "(Scholar Akison).
>
> *"In my view, I believe that the institution itself could work with CAPES to increase the number of scholarships for students,*
>
> *because it's a way for them to commit more to the course. " (Scholar Francisco).*
>
> *"In my opinion, the 'Pibidiano' should have more contact with the students at school and not just on the days when they are carrying out the projects. We could also be in the classroom helping students who have more difficulty assimilating the subjects taught by the teacher. "* (Maria).
>
> *"The partnership between scholarship holders and the whole school community needs to be improved, I notice a certain distance. "* (Scholar Luísa).
>
> *"Looking for mechanisms to make the student more present in the classroom, not just during interdisciplinary projects. In addition, increase resources for the purchase of materials to carry out projects. " (Fellow Very).*
>
> *'Becoming increasingly involved in interdisciplinary activities aimed at the whole school community. And in return, more frequent monitoring of students during lessons. "* (Maria Valentina).
>
> *"Increasing the number of partner schools within the city of Floriano/PI, so that students, supervisors and coordinators can attend to the largest number. " (Scholar João).*

The relationship between students in training and experienced teachers places the teaching-learning process in a perspective of knowledge construction. This dialogue, through situations, experiences and experiences in the classroom, allows for the construction of knowledge linked to teaching. In this sense, the teacher is not only a practitioner, but also responsible for the education of their students (Tardif, 2002). For this reason, PIBID has this bias of sharing knowledge through teaching practice, in which planning, integrating activities, the relationship between teacher and students and practical experience in the classroom are evident, awakening in the scholarship holder the desire to remain in teaching (Fejolo, Passos & Arruda, 2017).

FINAL CONSIDERATIONS

The Institutional Teaching Initiation Scholarship Programme (PIBID) is a teacher training programme that aims to introduce undergraduates to their future place of work, arousing interest in remaining in the teaching profession. In this way, the students on the Biological Sciences course at IFPI-Campus Floriano- PI, in their testimonies, stated that in the initial modules of the course they had difficulties in terms of fulfilling the activities, as they required, among other things, more orality/expressiveness skills and the exposition of arguments, postures, etc.

necessary for the teaching profession, as well as on methodological issues such as how to teach certain content, but after joining the programme, they had direct contact with the classroom, experienced the problems inherent in the teaching profession and other learning, which they would only have the opportunity to witness after completing the compulsory curricular internships.

The scholarship holders consider the programme to be an excellent professional qualification policy, since it values the teaching profession, the permanent dialogue between theory and practice, innovative teaching proposals through interdisciplinary projects, as well as dealing with conflicting situations in the classroom. Certainly, learning of this nature contributes to an understanding of what it means to be a teacher. With regard to the way the programme works, many of the fellows advocate greater involvement from partner schools, professionals in general and teachers, in order to promote integration, collective, collaborative work and strengthen teacher training. Therefore, it was found that the programme is of fundamental importance for the initial training of Biological Sciences undergraduates, as PIBID fosters a space conducive to the construction of knowledge in the educational environment.

REFERENCES

Abreu, M. F. (2012). The *contributions of the Institutional Teaching Initiation Scholarship Programme (Pibid) in the view of the participating school.* Retrieved from http://www.editorarealize.com.br/revistas/fiped/trabalhos/95192c98732387165bf8e 396c0f2dad2.pdf

Araújo, M. S., Souza, K. C., Sena, D. S., & Xavier, S. C. M. (2015). Experience Report of PIBID/ IFPI Campus Floriano, Subproject of Biology Reality of Scholarship Holders Facing Basic Education. In: *V Seminário Baiano Das Licenciaturas, V Seminário Estadual PIBID-IATe I Seminário PIBID do Nordeste.* Salvador, BA, Brazil.

., Sousa, C. P., &Cruz, E. R. (2016). The use of educational software as a technological didactic resource in the teaching of natural selection: an experience with biological sciences undergraduates at IFPI - Campus Floriano. In: *National Congress of Education (CONEDU).* Rio Grande do Norte,

RN, Brazil.

Arroyo, M. (2005). *New Configurations in the EJA Field.* In: Soares, L., Giovanetti, M. A., Gomes, N. L. (Orgs.). Dialogues in Youth and Adult Education. Belo Horizonte: Autêntica.

Bardin, L. (2011). *Content analysis.* São Paulo: Edições 70.

Barreiro, I. M. F., Gebran, R. A. (20060. *Teaching practice and supervised internship.* São Paulo: Avercamp.

Bizzo, N. (2009). *Science* - easy or difficult? São Paulo: Biruta.

Brazil (2002). Ministry of Education. National Education Council. Parecer CNE/ CP 009/2001. Diretrizes Curriculares Nacionais para a Formação de Professores da Educação Básica, em nível superior, curso de licenciatura, de graduação plena. *Federal Official Gazette,* Brasília, DF.

. (2010). Decree no. 7.219, of 24 June 2010. Provides for the Institutional Teaching Initiation Scholarship Programme - PIBID and makes other provisions. *Federal Official Gazette,* Executive Branch, Brasília, DF.

. (2012). *2009-2011 Management Report of the Basic Education Teacher Training Directorate.* Brasilia, DF. Retrieved from https://www.capes.gov.br/images/stories/download/bolsas/2562014-relatrorio-DEB-2013-web.pdf

. (2015). Ministry of Education.LDB *Lei de Diretrizes e Bases da Educação.* 11. ed. Brasília: Câmara.

. (2016). Ministry of Education.Le/ *de Diretrizes e Bases da Educação Nacional:* Lei n° 9.394, *de* 20 *de* dezembro *de* 1996.10. ed. Brasília: Câmara.

. (2016). Ministry of Health. *Resolution no. 510, of 07 April 2016.* Published in DOU no. 98, Tuesday 24 May 2016, section 1, pages 44, 45, 46.

. Ministry of Education (2013). *Regulations for the Institutional Teaching Initiation Scholarship Programme.* Brasília: Capes. Retrieved from http://www.capes.gov.br/images/stories/download/legislacao/Portaria 096 18jul13 ApproveRequallingPIBID.pdf

Cardoso, M. U. D. C., Araújo, M. S., Sousa, S. C. (2016). Expectations *versus* reality: a portrait of the pedagogical practice of teachers of youth and adult education in Floriano/PL In: *Congresso Nacional de Educação (CONEDU).* Rio Grande do Norte, RN, Brazil.

Carvalho, A. M. P., & Gil-Pérez, D. *Formação de professores de ciências:* tendências e inovações. 6. ed. São Paulo: Cortez, 2011.

Chaluh, L. N., Martins, B. S., Azevedo, M. A. R., & Osti, A. (2017). Supervisors' knowledge in the context of PIBID: training and partnership. *Comunicações Piracicaba,* 24(1), 125-147. Retrieved from https://www.metodista.br/revistas/revistas-unimep/index.php/comunicacoes/article/viewFile/2675/1924

Demo, P. (2004). *Teachers of the future and the reconstruction of knowledge.* Petrópolis: Vozes.

Dewey, J. (1958). *Philosophy in Reconstruction.* São Paulo: National Publishing Company.

Dornelesi, A. M., Galiazziii, M. C., & Altenhofeniii, S. (2016). Classroom stories in the academic and professional training of teachers in PIBID/FURG. *Revista Interinstitucional Artes de Educar,* 2(3), 100-116. Retrieved from www.e-publicacoes.ueri.br/index.php/riae/article/view/25697/19505

Fejolo, T. B., Passos, M. M., & Arruda, S. M. (2017). The socialisation of teaching knowledge: communication and professional training in the context of PIBID/Physics. *Investigations in Science Teaching,* 22(1), 103-126. Retrieved from https://www.if.ufrqs.br/cref/ols/index.php/ienci/article/download/390/pdf

Felício, H. M. S. (2014). PIBID as a "third space" for initial teacher training.Rewsfa *Diálogo Educação,* 14(42), 415-434. Retrieved from http://www2.pucpr.br/reol/index.php/dialoqo?dd99=pdf&dd1=12752

Freire, P. (1998). *Pedagogy of autonomy:* knowledge necessary for educational practice. 7. ed. Rio de Janeiro: Paz e Terra.

. (2011)..43. ed. São Paulo: Paz e Terra.

Frison, L. M. B., Veiga Simão, A. M., & Cigales, J. R. (2017). Teaching apprenticeships: PIBID and teacher training. *Revista e-Currícuium,* 15(1), 25-44. Retrieved from https://revistas.pucsp.br/index.php/curriculum/article/view/24263/22376

Garcia, C. M. (1999). *Teacher training* .for educational change. Porto: Porto Editora.

Gatti, B. A., André, M. E. D. A., Gimenes, N. A. S., & Ferragut, L. (2014). *An evaluative study of the Institutional Teaching Initiation Scholarship Programme (PIBID).* São Paulo: FCC/SEP.

Geglio, P. C., & Silva, A. F. (2015). The contributions and limits of PIBID as a government policy for teacher training. *Cadernos Cenpec\ New series,* 4(2), 95-107. Retrieved from http://cadernos.cenpec.orq.br/cadernos/index.php/cadernos/article/viewFile/290/28 7

Giacomelli, W., Borges, G. R., & Santos, E. G. (2016). Determinants of Demotivation at Work: A Theoretical and Empirical Investigation. *Revista de Administração de Roraima-UFRR,* 6(1), 4-17. Retrieved from http://revista.ufrr.br/adminrr/article/view/2602

Heidemann, L. A., & Oliveira, Â. M. M. Online tools in science teaching: a proposal with *Google Does. Revista Física na Escola,* 111(2), 30-33. Retrieved from http://www.sbfisica.orq.br/fne/Vol11/Num2/a09.pdf

Hemielewski, D. M. S., & Canton, V. D. (2014). PIBID in the field school: the supervisor's view of the teacher training process. In *Anais VI Fórum Internacional de Pedagogia.* Santa Maria, Rio Grande do Sul, Brazil. Retrieved from http://editorarealize.com.br/revistas/fiped/anais.php

Leal, C. L. C.; & Cardoso, E. S. (2015). Contributions to the Analysis of the Working Conditions and Health of Public Elementary School Geography Teachers in Santa Maria, RS*. *Revista Formação,* 1(22), 156-175. Retrieved from http://revista.fct.unesp.br/index.php/formacao/article/viewFile/3103/2944

Libâneo, J. C. (1994). *Didática.* 21. ed. São Paulo: Cortez editora.

.,& Pimenta, S. G. (1999). Training education professionals: a critical view and a perspective for change. *Educação & Sociedade,* 20(68), 239-277. Retrieved from http://www.scielo.br/pdf/es/v20n68/a13v2068.pdf

Lima, G. H.; Silva, R. S.; Arandas, M. J. G.; Lima Júnior, N. B.; Cândido, J. H. B., & Santos, K. R. P.

(2016). The Use of Practical Activities in Science Teaching in Public Schools in the Municipality of Vitória de Santo Antão-PE. *Rev. Ciênc. Ext,* 12(1), 19-27. Retrieved from http://ojs.unesp.br/index.php/revista proex/article/view/1190

Machado, J. C., Vasconcelos, M. C. C., & Oliveira, N. R. (2017). Initial and continuing teacher training: between discourse and practice. *Cad. Ed. Tec. Soc.,* 10(1), 13-27. Retrieved from http://www.cadernosets.com.br/index.php/cadernosets/article/view/409/199

Maldaner, O. A. (2007). *Situações de estudo no ensino médio:* nova compreensão de educação básica. São Paulo: Escrituras Editora.

Massena, E. P., & Siqueira, M. (2016). Contributions of PIBID to the Initial Training of Science Teachers from the Perspective of Undergraduates. *Brazilian Journal of Research in Science Education,* 16(1), 17-34. Retrieved from https://seer.ufmg.br/index.php/rbpec/article/viewFile/2539/1940

Moran, J. M. (2004). The new spaces in which teachers work with technology. *Revista Diálogo Educacional, Curitiba,* 4(12), 13-21. Retrieved from http://www.eca.usp.br/prof/moran/site/textos/tecnologias eduacacao/espacos.pdf

Nóvoa, N. (1991). *"Teachers: in search of a lost autonomy?".* In Educational Sciences in Portugal - Current situation and perspectives. Porto: SPCE.

Oliveira, C. L. (2006). *Meaning and contributions of affectivity in the context of Project Methodology in Basic Education.* 2006. Dissertation (Master's Degree in Technological Education) - CEFET-MG, Belo Horizonte-MG.

Paranhos, M. L. R., Souza Filho, L. C., & Paranhos, M. C. R. (2016). Reflections on the teaching profession: building experiences from observations in science classes. *Journal of the Brazilian Society of Biology Teaching,* 6(9), 5752- 5761. Retrieved from http://www.sbenbio.org.br/wordpress/wp- content/uploads/renbio-9/pdfs/2426.pdf

Pimenta, S. G. (2012). *Pedagogical knowledge and teaching activity.* 8.ed. São Paulo: Cortez.

Ramos, D. C. P.; Araújo, R. S.; & Santana, R. O. (2016). The Internet as a source of analysis for PIBID and initial teacher training. *Revista de Pesquisa Interdisciplinar,* Cajazeiras, v. 1, ed. Especial, 63-72. Retrieved from http://revistas.ufcg.edu.br/cfp/index.php/pesquisainterdisciplinar/article/download/6 9/47

Ribeiro Júnior, J. I. (2013). *Statistical analysis in Excel: a* practical guide. 2. ed. Viçosa, MG: UFV.

Ribeiro, S. S. (2013). *Graduate students' perceptions of the contributions of PIBID - Maths.* Dissertation (Professional Master's Degree in Education) - Federal University of Lavras, Lavras.

Sallai, R. A. (2016). Importance of PIBID in initial teacher training at the Santo André Foundation University Centre. *Educação & Linguagem,* 19(1), 79- 96. Retrieved from https://www.metodista.br/revistas/revistas- ims/index.php/EL/article/view/7097/5433

Saviani, D. (2009). Teacher training: historical and theoretical aspects of the problem in the Brazilian context. *Revista Brasileira de Educação,* 14(40), 143-155. Retrieved from http://www.scielo.br/pdf/rbedu/v14n40/v14n40a12.pdf

Schõn, D. A. (2000). *Educating the reflective professional:* a design for teaching and learning. Porto

Alegre: Artmed.

Tardif, M. (2002). *Teaching knowledge and professional training.* Petrópolis: Vozes.

., Borges, C., & Maio, A. (2012). *Le virage réflexifen éducation:* ou en sommes-nous 30 ans après Schõn? Brussels: Editora de Boeck Supérieur.

Viana, B. O. S., Almeida, O. S.; Freitas, M. S.; & Pereira, N. A. (2016). Impacts of PIBID on the training of Biological Sciences undergraduates at UESB: a

experience report. *Revista Ensino & Pesquisa,* 14(1), 260-276. Retrieved from http://periodicos.unespar.edu.br/index.php/ensinoepesquisa/article/view/864/538

. (2016). Impacts of PIBID on the training of biological sciences undergraduates at UESB: an experience report. *Revista Ensino & Pesquisa,* 14(1), jan./jun. 7543 p.

Zabala, A. (2008). *Educational practice - how to teach.* Porto Alegre: Artmed.

Zeichner, K. M. A. (1993). *Reflective teacher education:* ideas and practices. Lisbon: Educa.

CHAPTER 2

PRACTICAL LESSONS IN THE CONSTRUCTION OF SCIENTIFIC KNOWLEDGE IN BIOLOGY: A REPORT FROM STUDENTS AT A PUBLIC SCHOOL IN FLORIANO/PI

Maurício dos Santos Araújo [mauriciosangesl 1@hotmail.com]

Master's student in Genetics and Improvement at the Federal University of Piauí. He is currently working on Quantitative Genetics (genetic improvement of cowpea) and Toxicological Genetics with toxic, cytotoxic, genotoxic and mutagenic evaluations of medicinal plants. He develops projects aimed at teaching Genetics, with an emphasis on the construction of didactic-pedagogical materials.

Abstract: Experimentation in the teaching of Biology emerged around the 19th century with the aim of uniting theory and practice. The aim of this article is to report on the results obtained from the project entitled *"Practical classes in the Biology laboratory"*, developed at a secondary vocational school integrated with the technical course in Clinical Analyses in Floriano/PI, as well as to identify the contributions made by this activity to the students' professional training. The research followed a qualitative approach with a descriptive focus, in which 14 (fourteen) secondary school students took part.[a] . The data collection instrument was two questionnaires, one diagnostic and the other reflective *(pre-* and post-test), and *on-site* observations. A practical laboratory lesson was carried out as an intervention activity in the Microscopy laboratory at IFPI-Campus Floriano, after identifying the difficulties faced by the students in some of the content required for their professional training. It was found that 71.42% of the students did not take practical classes during the technical course, a factor which can hinder the construction of knowledge. With regard to practical classes, 57.14% said they were unprepared to enter the labour market based on the knowledge they were acquiring, as they commented that it was insufficient for them to be able to work hard in their future workplace, i.e. in the laboratory. The project therefore gave the students direct contact with the laboratory environment, making it a constructive tool in their professional training. In addition, they were able to recall basic concepts relating to cell biology, biochemistry and the rules governing the laboratory environment.

Keywords: Practical lessons. Experimentation. Professional training. Theory and practice.

1 INTRODUCTION

Biology teaching has undergone many transformations over the years. The active participation of students in practical activities in the classroom is becoming more difficult every day, a factor that can often hinder the realisation of pedagogical practice. The students' lack of knowledge about certain biological processes covered in class is a problem faced by many teachers when teaching biology. In this sense, practice through experimentation is one of the mechanisms used to demonstrate the importance of these processes (LIMA et aL, 2016).

Carrying out experiments as a pedagogical tool in biology teaching can contribute to building meaningful and integrative learning, as it is possible to visualise certain events in the student's daily life through these processes, as well as being able to correlate the subjects worked on in the classroom (ALBUQUERQUE et aL, 2014). From this perspective, experimentation not only aims to present certain phenomena that are present in the student's context, but also to develop the ability to define hypotheses, understand the intrinsic aspects of the problem encountered, understand the process of simplifying and modelling problems, formulate and test hypotheses and draw up the results found (TAKAHASHI; CARDOSO, 2012).

In view of the process of experimentation in biology teaching, laboratory practice arises with the aim of demonstrating how certain biological processes that permeate the area of Natural Sciences work. It can also encourage students to develop scientific and investigative thinking, making them central agents in the teaching and learning process (MEIRE et al., 2016). However, laboratory practice cannot replace the theory worked on in the classroom, but it does provide the necessary conditions for students to plan, act and understand how the processes around them happen. In this way, experimentation acts as a basic premise for the researched investigation (LIMA; TEIXEIRA, 2014).

According to the National Curriculum Parameters (PCNs), understanding the basic processes of biology provides an understanding of the living world and the relationship between human beings and other organisms, all as a function of their relationship with the environment in which they live (BRASIL, 1999). The use of new alternative methodologies in the teaching of natural sciences, such as practical laboratory lessons, puts the student in a new perspective. This practical model seeks to detach itself from the requirements left by the traditional current that for decades placed the student in a perspective of mere receiver of information, making them a passive agent in the teaching-learning process and the teacher assuming in this context a posture of holder of knowledge (BICHO; QUEIROZ; RAMOS, 2016).

The Law of Guidelines and Bases of Education (LDB) No. 9.394/96, of 20 December 1996 in its article 35, item IV states that "it is essential to understand the scientific-technological foundations of production processes, relating theory to practice, in the teaching of each discipline. " (BRASIL, 2015, p. 47). Therefore, it is possible for students to develop all this scientific-technological apparatus with regard to Biology, as laboratory practice combined with theoretical knowledge can provide students with the possibility of coming into contact with biological phenomena, equipment, processes and routines, associating theory and practice so that they can apply this knowledge in their daily lives (FULAN et al., 2014; KRASILCHIK, 2016).

The search for short-term, quality vocational training that provides the minimum conditions necessary to compete in the labour market has become a constant quest for students (OLIVEIRA; ESCOTT, 2015). Over the years, technical courses in the natural sciences have suffered from a lack of physical space, reagents and intellectual labour as one of the main factors contributing to school dropout (ARAÚJO; SOUSA; SOUSA, 2016). For this reason, the lack of resources for teaching science and biology is a reality faced by many Brazilian schools today.

Classes with a different profile have been gaining ground with the training of new professionals, who realise that traditional education no longer has the same reach as it did decades ago. Teachers have sought to bring the content covered into the students' reality, which happens more efficiently when practical lessons are carried out (SILVA et aL, 2014). In the teaching-learning process, alternative tools that provide the teacher with the necessary means to carry out a dynamic lesson with a prominent rate of knowledge exchange deserve recognition for their value in student learning. New teaching methodologies have shown that there is a disparity in how students receive certain content (SANCHES; RAMOS; COSTA, 2014).

Actions that relate theory and practice, which lead students to self-reflect on the subjects worked on in the classroom, enable them to self-assess how their learning is being conceived (VIANA, 2016). In this sense, the teacher can use alternative methodologies to make the teaching-learning process more consistent with the students' reality, leading them to associate the theory worked on in the classroom with the practice carried out (SANTOS et aL, 2013). Many Brazilian schools don't have specific laboratories to carry out certain practices, but this factor can't prevent teachers from carrying out these activities, as there are countless easy experiments that can be quickly assimilated by students; all that's needed is for the teacher to analyse the possibilities and resources available at the school (PAGEL; CAMPOS; BATITUCCI, 2015).

The laboratory has a motivating aspect in the teaching-learning process, as it provides students with various experiences in terms of practical lessons. It also helps them to remember certain content and put it into practice (DOHN et al., 2016). Experimentation within a laboratory space emerged around the 19th century, seeking to place the student at the centre of the teaching-learning process, applying the concepts constructed during their student cycle to the realisation of knowledge (MORAIS; SANTOS, 2016).

Initial training is the period in which the academic is building the beginnings of being a teacher. Therefore, the use of new teaching methodologies helps them to build an educating attitude and the ability to develop their pedagogical practice geared towards innovation in the classroom (RAMOS; ARAÚJO; SANTANA, 2016; ARAÚJO; SOUSA; LEITE, 2017). Therefore, practice in biology teaching is inseparable from theory, as it contributes to the construction of knowledge and promotes reflection on the conceptual apparatus built up during the student cycle (OLIVEIRA; SCHNEIDER, 2016).

From time to time, Brazilian public education presents characteristics that make up the profile of the new teaching models, but it is notable that the attachment to traditional tools is still very strong. This is noticeable in assessment, in monitoring students, and in the role of the teacher. Little by little, the school environment has been moulding itself to the needs of students in a slow process that still needs some time to meet all the demands. Practical lessons in the laboratory have emerged as an opportunity to provide students with knowledge that the theoretical approach is unable to generate; understanding how certain processes happen is essential if learning is to be truly built into the student's educational trajectory, especially in secondary school. Identifying how high school students are receiving laboratory practices and what the repercussions of this knowledge are on their learning process is therefore essential.

The aim of this article is to report on the results obtained from the project entitled *"Practical classes in the Biology laboratory"*, developed at a secondary vocational school integrated with the technical course in Clinical Analyses in Floriano/PI.

2 METHODOLOGY

The theoretical and methodological bases of this study are based on a qualitative perspective with a described approach. The work was developed through the application of a project entitled *"Practical classes in the Biology laboratory"*, which was carried out at the end of the first semester of 2016 and concluded at the beginning of the first semester of 2017, with students on the technical course in Clinical Analyses integrated into secondary education. [a]The sample group was 14 (fourteen) secondary school students from a technical school in the city of Floriano/PI (Figure 1). It is worth emphasising that in order to preserve the identity of the subjects, participants in this investigation, fictitious names/codenames were used to give meaning to the speeches of the research participants and make reading the text more enjoyable for the reader.

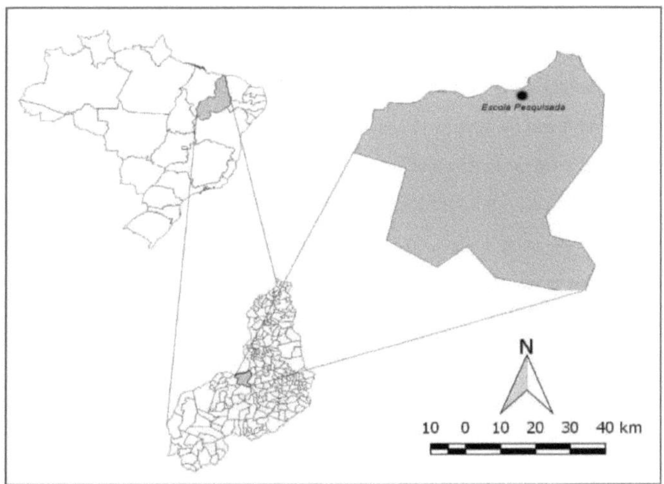

Figura 1. Map showing the location of the research school in the city of Floriano/PI.
Source: Empirical Research Data (2018)

Questionnaires containing open and closed questions were used as a data collection tool to find out about the difficulties, expectations and reality of the course from the students' perspective. *In* addition, during the methodological process, *on-site* observations were also used as a collection tool. Initially, the questionnaire (pre-test) was administered to the students, who took part in the research, with the aim of identifying how the association between theory and practice is being made during the course, as well as the frequency with which practical lessons are being carried out by the Biology teacher. After tabulating and descriptively analysing the data (pre-test), a partnership was set up with the Federal Institute of Education, Science and Technology of Piauí - IFPI, Floriano *Campus*, in order to provide the students with a space to carry out a laboratory lesson, with the aim of contributing to the construction of scientific knowledge.

After the practical lesson in the laboratory, the second questionnaire (post-test) was administered to the students who had taken part in the practical. The aim of this questionnaire was to identify the contributions of laboratory practice to the construction of professional identity.

The practical class was held in the Microscopy Laboratory of the *\FP\-Campus* Floriano with the aim of extracting Deoxyribonucleic Acid (DNA) from strawberries, following as a theoretical basis the experimental protocol of Beluzzo, Oyakawa and Bueno (2016) with adaptations by Sousa et al. (2016). At first, the following materials were used: 03 (three) ripe strawberries, distilled water, 98% commercial alcohol, sodium chloride (NaCl), transparent plastic bags, beaker, colourless detergent, gauze for filtering, beakers, sieve, test tubes, measuring spoon, glass rods and funnel.

The practice was carried out as follows: the students were grouped into three (03) groups.

Afterwards, the experimental protocol containing the procedures for the operationalisation of the lesson was read out to them and then they took the three ripe strawberries and separated the fruit from the leaves. The sample was placed in a transparent plastic bag and macerated until homogenised.

The hydrolysis solution was made by adding 150 ml of distilled water, a tablespoon of colourless detergent, a teaspoon of NaCL. The solution was mixed with a glass rod and then 1/3 of the hydrolysis solution was added to the mixture containing the strawberry, waiting 30 minutes until it was completely homogenised. The solution was sieved using a sieve to remove the sediment it contained. Finally, the solution was transferred to the test tubes using a funnel and ice-cold 98% ethyl alcohol was added until the genetic material precipitated.

The results described in the questionnaires *(pre-* and post-test) were tabulated and analysed using the *Statistical Package for the Social Sciences (SPSS)* as a statistical parameter.

3 RESULTS AND DISCUSSION

The research was carried out with students on the technical course in Clinical Analyses at a school in Floriano/PI. To carry out the work, the project entitled *"Practical classes in the Biology laboratory"* was applied. With regard to the gender of those interviewed, 42.9% were male and 57.1% female, with an average age of 16±1 years, as illustrated in (Table 1).

Table 1. Distribution of students at the Clinical Analyses Technical School by gender.

Gender	Frequency	Percentage (%)
Female	8	57,1
Male	6	42,9
Total	14	100

Source: Research data (2017).

Practical lessons in biology teaching have attracted the attention of many education professionals for their ability to provide the ideal space for linking theory and practice through simple and easy procedures that can be worked on in the classroom. With this in mind, we sought to find out from those surveyed how often the teacher carried out practical lessons in the subject of Biology. It was found that 71.42% of the students didn't take practical classes in the Biology laboratory because the school is undergoing renovations to its physical space and the work has not yet been completed, a factor which is detrimental to the quality of the course.

In this sense, 14.29% of the students said that practical classes are often held every fortnight when the school manages to sign partnerships with other educational institutions, and that there is no regularity to this activity for all students. In addition, 14.29% of the students said that their teacher carried out this activity sporadically, a fact that gives practical lessons even greater importance given that the region in which they live, the labour market demands professionals with extensive experience in the specific area, as illustrated in (Figure 2).

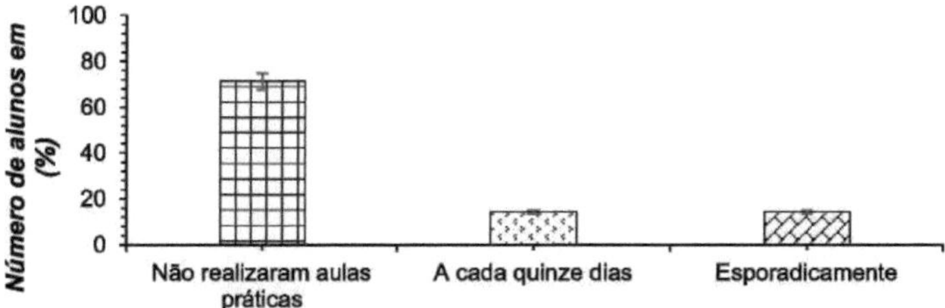

Figura 2. Frequency with which students on the Clinical Analyses technical course take practical classes in the laboratory.

Source: Research data (2017).

Carrying out experimental activities takes on a new meaning in being and doing teaching. Through this activity it is possible for students to interact with others in order to solve certain problems. Joining forces and working to develop and build integrative learning (KRASILCHICK, 2016). Corroborating this idea, Silva and Serra (2013) discuss practical classes and emphasise their importance for teaching biology, as they contribute to the construction of meaningful knowledge and also act in motivating students for the processes inherent in the subject.

When the students were asked about the irregularity of practical classes among them, they discussed carrying out this activity through internships, *"as we don't have practical classes here at school, we often get internships and carry out these practices. "* (Maria Alice). In addition, another student referred to the partnerships that some teachers had with other educational institutions, reporting that: *"some of our teachers get the laboratory of other institutions, such as the Federal University of Piauí (UFPI), State University of Piauí (UESPI) and the Federal Institute of Piauí (IFPI), to carry out these practices, and often the space is small for the amount of students. "* (Student Bruna Ananda).

Activities that stimulate the development of students' abilities by relating theory and practice make learning more meaningful, providing moments of challenge and investigation, awakening in students a critical and investigative sense (BIZZO, 2001; SOARES; BAIOTTO, 2015; ARAÚJO etaL, 2018). In this way, the teacher is an instrument of great importance in this context. Through a humanistic pedagogical practice, they can help students understand certain aspects of biology through experimentation, and by carrying out experiments, they can correlate them with the content being taught in class.

Courses in the natural sciences should be intrinsically linked to practice in order to bring the conceptual apparatus closer to the students' reality. Based on this understanding, the students were asked how this association between theory and practice is going: *"I think my training is being interrupted because I can't put what I'm studying into practice. " (Student Maria).* In this sense, Piaget (1981, p.82) states that "developed concrete operations allow for exact observation and experimentation". Thus, specifically in the teaching of biology, many processes need to be related to practice, so that the student can more easily understand what is being worked on.

Many students ask themselves about their training for the labour market. In this sense, when asked about their preparation to enter the labour market. One student stated that, *"I don't feel prepared to work in a Clinical Analyses laboratory, because the conceptual part worked on in the classroom is insufficient and we don't have basic laboratory practices to work in the area, one of the reasons why most students drop out of the course. " (Student Lorenzo).* In this sense, carrying out practical activities makes it possible to gain a greater understanding of the processes studied, especially with regard to the teaching of Science and Biology, and is a highly relevant tool for their training (SILVA et aL, 2016).

After the practical class in the microscopy laboratory, we tried to identify the contributions of this activity to the students' professional training. In this way, 64.28% said that, through the practical class, they gained a better understanding of the laboratory routine. Around 21.43% said they had been able to understand the experimental activity step by step and 14.29% said they had acquired a broader knowledge of the equipment and glassware in the laboratory (Figure 3).

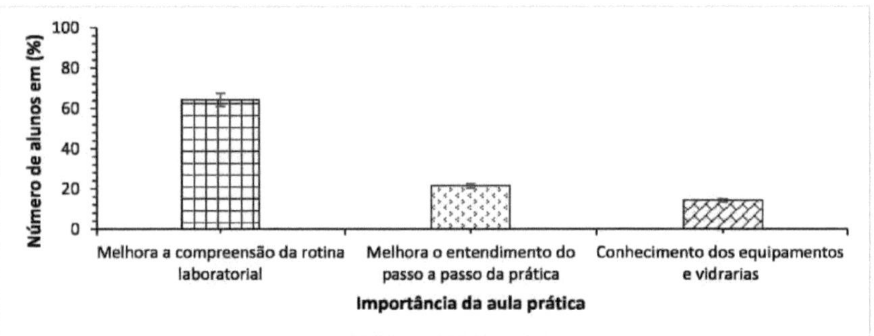

Figura 3. Statements by the students of the Clinical Analyses technical course who took the practical class in the laboratory.

Source: Research data (2017).

According to Poletti (2001), carrying out experimental activities is indispensable in the school context because they must be directly related to theory. It is through this that students can strengthen the knowledge they have built up in the classroom, concretising this knowledge. For this reason, carrying out this activity in science and biology teaching provides students with an understanding of the laboratory routine, from the step-by-step practice to the materials that are used, contributing to a broader understanding of the processes that surround the educational field, contributing to the construction of learning in a holistic way (VIEIRA et aL, 2013).Entering the labour market is a time of great change in a student's life. When students were asked about this, 14.29% reported that they were prepared to enter the labour market with the training they were receiving during their studies. However, 21.43% of those interviewed said that their training was incomplete, 7.14% said that the lack of practice was one of the main problems they faced at school and 57.14% of the students said that they were unprepared to enter the labour market due to the lack of this activity in their professional training. This data is statistically corroborated by the correlation with the frequency with which students take practical classes ($p < 0.01$) (Figure 4).

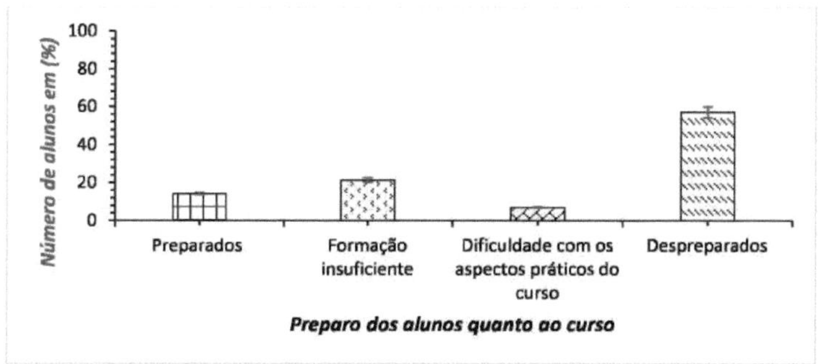

Figura 4. Statements by students on the Clinical Analyses technical course about their expectations for their future profession.

Source: Research data (2017).

According to LDB No. 9.394/96 in Section IV of High School in its Art. 35, "the basic preparation for work and citizenship of the student, to continue learning, so as to be able to adapt flexibly to new conditions of occupation or further training" (BRASIL, 2015, p. 24). That's why the school, together with the family and the state, must provide the minimum conditions for students to prepare to enter the labour market. In addition, it is a critical period that permeates the student's life, as they depend on the support of their family, school and encouragement from friends, to try to reconcile academic life with work (SARRIERA; VERDIN, 1996).

4 FINAL CONSIDERATIONS

Complementary activities in science and biology teaching have made many contributions to students' learning in the classroom. Practical laboratory lessons are one of the tools used to bring theory and practice closer together. The project provided students on the Clinical Analyses technical course with direct contact with the laboratory environment through the chosen practice, a constructivist tool in professional training, since this activity was not found to be carried out regularly as a teaching method at the school studied. It was therefore found that the school does not have a space ready for practical classes, which hinders the students' training process and distances them from the professional reality.

In the educational sphere, the lack of resources and especially of appropriate spaces for carrying out practical activities has always been under discussion. The lack of laboratories for students to carry out certain experiments is one of the problems faced during professional training. It is therefore necessary for schools that do not have this space for practical lessons to establish partnerships with higher education institutions as an alternative to contribute to the training of these students. In addition, there are countless practices that don't need to be carried out only in a modernised laboratory. All that is needed is for the teacher to look for simple, objective experiments that can be adapted to the reality of the classroom, as a way of minimising this problem that persists during the process of training these students.

REFERENCES

ALBUQUERQUE, F. P. et al. Entomology in agricultural technical secondary education: A working proposal.

Revista Eletrónica de Educação, v. 8, n. 3, p. 251-265, 2014.

ARAÚJO, M. S.; SOUSA, S. C.; LEITE, A. S. Active methodologies for the inclusion of hearing-impaired students: new possibilities in the teaching-learning process in biology. In: CARLONI, P. R.; FREIRE, A. C.; ANDRADE, T. C. O. **Inclusão, Educação e Sociedade.** Goiânia: Mundial Gráfica, 2017.

ARAÚJO, M. S.; SOUSA, C. P.; SOUSA, S. C. Aulas práticas em laboratório: relato de discentes do curso técnico em análises clínicas de uma escola pública em Floriano/PJ. II ENCONTRO NACIONAL DE PESQUISAS E PRÁTICAS EM EDUCAÇÃO - RN, Rio Grande do Norte: 2016. **Proceedings...** Rio Grande do Norte: UFRN, 2016.

ARAÚJO, M. S. et al. Genetics in the classroom context: difficulties and challenges in a public school in Floriano-PI. REnCiMa, v. 9, n. 1, p. 19-30, 2018. BELUZZO, E. M.; OYAKAWA, J.; BUENO, R. S. **Homemade strawberry DNA extraction.** São Paulo: USP, 2016.

BICHO, V. A.; QUEIROZ, L. C. S.; RAMOS, G. C. Experimentation in youth and adult education: a meaningful practice in the teaching-learning process.
Revista Scientia Plena, v.12, n. 12, p. 1-8, 2016.

BIZZO, N. **Ciências.** 2. ed. São Paulo: Ática, 2001.

BRAZIL. Ministry of Education. **LDB Law of Guidelines and Bases of National Education.** 11. ed. Brasília: Câmara, 2015.

. Ministry of Education. **National Curriculum Parameters:** Secondary Education. Brasília: MEC/SEF, 1999.

DOHN, N. B. etal. Students' motivation towards laboratory work in physiology teaching. **Adv Physiol Educ,** v. 40, n. 3, p. 313-328, 2016.

FULAN, J. A. et al. The application of practical classes in the teaching of science and biology in the municipality of Humaitá-AM. **Revista Simbio-Logias,** v. 7, n. 10, p. 16-23, 2014.

KRASILCHICK, M. **Biology teaching practice.** 4. ed. São Paulo: Edusp, 2016.

LIMA, G. H. et al. The Use of Practical Activities in Science Teaching in Public Schools in the Municipality of Vitória de Santo Antão-PE. **Rev. Ciênc. Ext,** v. 12, n. 1, p. 19-27, 2016.

LIMA, K. E. C.; TEIXEIRA, F. M. Experimentation in science teaching for the appropriation of scientific knowledge. **Revista da SBEnBIO,** v. 1, n. 7, p. 4516- 4527, 2014.

MEIRE, I. A. et al. Teaching-Learning through Microbiology Laboratory Practices. **Science & Technology Magazine: FATEC-JB,** v.8, n.1, p.1-8, 2016.

MORAIS, V. C. S.; SANTOS, A. B. Implications of the use of experimental activities in the teaching of biology in public schools. **Revista Investigações em Ensino de Ciências,** v. 21, n. 1, p. 166-181, 2016.

OLIVEIRA, A. R. M.; ESCOTT, C. M. Políticas públicas e o ensino profissional no Brasil. **Revista Ensaio: aval. pol. publ. Educ,** v. 23, n. 88, p. 717-738, 2015.

OLIVEIRA, J. M. P.; SCHNEIDER, E. M. Work projects: an alternative in initial training for theoretical-practical articulation. **Revista de Educación en Biologia,** v. 19, n. 1, p. 19-34, 2016.

PAGEL, U. R.; CAMPOS, L. M.; BATITUCCI, M. C. P. Contribution of practical classes in the teaching-learning process of Biology. **Revista Experiências em Ensino de Ciências,** v. 10, n. 2, p. 14-25, 2015.

PIAGET, J. **Genetic Epistomology.** São Paulo: Martins Fontes, 1981.

POLETTI, N. **Estrutura e Funcionamento do Ensino Fundamental.** 26. ed. São Paulo: Ática, 2001.

RAMOS, D. C. P.; ARAÚJO, R. S.; SANTANA, R. O. The Internet as a source of analysis for PIBID and initial teacher training. **Revista de Pesquisa Interdisciplinar,** v. 1, n. 1, p. 63-72, 2016.

SANCHES, K.S; RAMOS, A.O.; COSTA, F.J. Digital technologies and the need for continued training of Science and Biology teachers for technology: a study carried out in a school in Belo Horizonte. **Revista Tecnologias na Educação,** v. 6, n. 11, p. 1-11,2014.

SANTOS, D. P. et al. Analysis of a practical lesson on DNA extraction from plant cells in a public school in Arapiraca. 65th Annual Meeting of the SBPC - AL, Alagoas: 2017. **Proceedings...** Alagoas: Brazilian Society for the Advancement of Science, 2017.

SARRIERA, J. C.; VERDIN, R. Young People Looking for Work: A Qualitative Analysis. **Revista PSICO,** v. 27, n. 1, p. 59-70, 1996.

SILVA, A. T. et al. Practical classes: their importance and effectiveness in teaching Biology. Revista Univap, v. 22, n. 40, p. 569-569, 2016.

SILVA, P.F.R.S., CAETANO, G.T.P., SILVA, A.P. The importance of practical classes in the teaching-learning process in primary school. In: V *Encontro Nacional das Licenciatura, IV Seminário Nacional do PIBID e XI Seminário de Iniciação a Docência da UFRN,* 2014, Natal. Teachers' training space. Natal, 2014. p. 1- 10.

SILVA, S. M.; SERRA, H. Investigation on experimental activities of physical knowledge in the early grades. **Revista Brasileira de Pesquisa em Educação em Ciências,** v. 13, n. 3, p. 9-23, 2013.

SOARES, R. M.; BAIOTTO, C. R. Biology practical classes: their applications and the counterpoint of this practice. **Revista Di@Logus,** v. 4, n. 3, p. 53-68, 2015.

SOUSA, I. C. et al. The Importance of the Practical Class in the Biology Laboratory: Formative Tool in the Teaching-Learning Process of Students of the Technical Course in Clinical Analyses in Floriano/PI. III NATIONAL CONGRESS OF EDUCATION (CONEDU) - RN, Rio Grande do Norte: 2017. **Proceedings...** Rio Grande do Norte: UFRN, 2016.

TAKAHASHI, E. K., CARDOSO, D. C. Remote experimentation in formal teaching activities: a study from Qualis A journals. **Revista Brasileira em Pesquisa em Educação em Ciências,** v. 11, n. 3, p. 85-208, 2012.

VIANA, B. O. S. Impacts of PIBID on the training of undergraduate students in biological sciences at UESB: an experience report. **Revista Ensino & Pesquisa,** v. 14, n. 1, p. 261-276, 2016.

VIEIRA, B. C. R. et al. The importance of experimentation in science for the construction of knowledge in primary education. **Revista Enciclopédia Biosfera,** v. 9, n. 16, p. 2276-2285, 2013.

CHAPTER 3

THE STUDY OF PALEONTOLOGY FOR UNDERSTANDING EVOLUTIONARY PROCESSES: PERCEPTIONS OF HIGH SCHOOL STUDENTS AT THE DR. PAULO RAMOS TEACHING CENTRE IN SÃO JOÃO DOS PATOS/MA, BRAZIL

Maurício dos Santos Araújo, Michelle Mara de Oliveira Lima, Willamys Rangel Nunes de Sousa, Aracelli de Sousa Leite, Samara Silva Siqueira

Federal Institute of Education, Science and Technology of Piauí - IFPI, Campus Floriano, Piauí, Brazil. E-mails: mauriciosangesl 1@hotmail.com. michellelima@ifDi.edu.br, ranaelnunes@gmail.com. aracellileite@ifDi.edu.br, samarasio@gmail.com

Abstract: The aim of this study was to analyse the perceptions of students from a public school in São João do Patos/MA, Brazil, regarding the study of fossils for understanding evolutionary changes in biological populations. [a]The research was carried out with 33 (thirty-three) high school students from the Centro de Ensino Dr. Paulo Ramos. A form with open and closed questions was used as a data collection tool to identify how the students perceived the dynamics of the Earth, with regard to the study of rocks and fossils for understanding evolutionary mechanisms. The data was analysed using criteria already published in the literature. It was observed that most of the students had a satisfactory conception of the importance of the fossil record for understanding the theory of biological evolution of species, showing that the biology teacher had worked on geological and palaeontological principles to facilitate understanding of evolution. It is therefore necessary to work on these concepts from Geology, Palaeontology and Evolution in an interdisciplinary way, so that students can build solid, integrated knowledge, thus contributing to the consolidation of meaningful learning.

Keywords: geology, evolution, fossil record, rocks, palaeontology.

INTRODUCTION

Palaeontology is a science whose object of study is fossil records. The word fossil has Latin origins meaning "unearthed" and includes any remains or preserved traces in which animals, plants or other living beings can be found that inhabited that space in a certain period (Sene, 2016). These records are direct, concrete evidence that proves the existence of the evolutionary history of living beings in different periods on planet Earth. For this reason, it is considered one of the areas of knowledge that makes it possible to elucidate evolutionary and temporal aspects, as well as the ability to understand the origin of planet Earth and its dynamics (Novais et al., 2015).

Most fossils are found in sedimentary rocks, which are formed by the deposition and solidification of sediments caused by weathering and modifying agents. However, species of mammoths have been identified in ice and insects in amber, where the specimen remained in a good state of preservation. Amber is a plant resin that preserves various organisms, such as pollens, cyanobacteria, beetles, mosquitoes, wasps and some small amphibians (Futuyma, 2009).

The fossil record is of great importance for understanding the relationships between organisms and the ecosystem (Muscente et al., 2017). They are considered evidence that living beings inhabited the region where they were found at a certain time. An example of a fossil is coprolites, which are the remains of faeces that have undergone a process of desiccation and mineralisation over the years (Sene, 2016). For this reason, there is an intrinsic relationship between Geology, Palaeontology and the Biological Sciences as areas of knowledge that seek to explain the mechanisms that contributed to the preservation of these palaeontological finds and the changes that occurred on the Earth's surface through geological mechanisms (Cassab, 2010; Novais et al., 2015).

Palaeontology plays an important role today, as it provides evidence of the existence of evolutionary mechanisms within a biological population (Mendes, 1986; Araújo Júnior and Porpino, 2010). Fossils serve as a 'record' that shows the transformations that organisms have undergone throughout history. In this way, the "(...) practical aim of palaeontology is to estimate the relative dating of the layers, by the degree of evolution or by the occurrence of various groups of fossil plants and animals." (Zucon, 2011, p.12). It is therefore possible to identify the rocks, remains of plants, animals, mineral substances and fuels such as phosphate, coal and oil that serve as the basis for the study of Geology, and there is an interdisciplinary relationship between Biological Sciences and Geology with the aim of studying the types of rocks in which the fossil record is deposited (Alonço and Boelter, 2016).

Biologists have already come to the conclusion that evolution is the central theme within biology. The evolutionary transformations identified in species through fossil records are evidenced through the evolutionary mechanisms that act on the genetic material of the species, which can present changes at the genotypic and phenotypic level (Reece et aL, 2015). The consensus among biologists is based on the evolutionary concept of Theodosius Dobzhansky (1900-1975) in which "nothing makes sense in Biology except in the light of Evolution." (Dobzhansky, 1973, p.125). Therefore, in order for the theory of Natural Selection proposed by Charles Darwin (1809-1882) and with the participation of his co-author Alfred Wallace (1823-1913) to be accepted by the *Linnean Society of Lodon* and the international scientific community of that time, it was necessary for the naturalist to look for evidence to prove the principles on which his theory of evolution was based.

As a result, he spent two decades searching for facts that would corroborate his ideals, looking at various areas of knowledge, including botany, zoology, embryology and palaeontology. Darwin's contributions to science were of great importance, as he explained in a detailed and systematised way the mechanisms of natural selection within nature and the overlap of certain species in certain *habitats* (Griffiths et al., 2016). One of the great legacies left by Darwin for geology and other sciences was the true age of planet Earth.

In a study described in Chile around 1835, it was observed that there was a certain elevation of the Andes Mountains, as it was identified in a petrified forest whose trunks resembled those of the araucarias found at sea level, found at a height of more than 2,000 metres. Marine fossils were found in layers of earth in the mountains, but it was identified that there was a 2 centimetre gap. These observations corroborated the idea that the Earth was older than previously thought (Grotzinger and Jordan, 2013). Darwin therefore contributed in the sense that, in order for species to evolve over the years in natural space, it was necessary for the Earth to have a long period of existence that would allow evolutionary mechanisms to act on populations. Such ideas contradicted the precepts preached by the Catholic Church, which claimed that the Earth was only 7,000 years old (Carvalho, 2010).

The geological time scale shows the division of the main geological and biological events that have taken place in the Earth's history over billions of years. These events can therefore be divided into eons, eras, periods and epochs. Most of the geological and biological events were ordered by Charles Darwin and some geologists based on the ideological conceptions of Charles Lyell, author of the work entitled *"Principles of Geology",* and many fossil records were identified in the Archaean era 3,600 million years ago (Table 1) (Futuyma, 2009).

Table 1. - The geological time scale with the main biological and geological events.

ERA	PERIOD	ÉPOCA	MILLIONS OF YEARS FROM THE BEGINNING TO THE PRESENT	MAIN EVENTS
CENOZOIC		Recent (Holocene)	0,01	Continents in modern positions; repeated glaciations and lowering of sea levels; shifts in geographical distributions; extinctions of large mammals and birds; - evolution from *Homo erectus* to *Homo sapiens;* emergence of agriculture and civilisations.
	Quaternary	Pleistocene	1,8	
	Tertiary	Pliocene	5,2	- Continents approaching their modern positions; climates drier and colder; radiation from mammals, birds, snakes, angiosperms, insects pollinators, teleost fish.
		Miocene	23,8	
		Oligocene	33,5	
		Eocene	55,6	
		Palaeocene	65	
MESOZOIC	Cretaceous		144	Most continents separated; radiation of dinosaurs; increased diversity of angiosperms, mammals, birds; mass extinction at the end of the period; including the last ammonites and dinosaurs.
	Jurassic		206	Continents separating; various dinosaurs and other "reptiles"; first birds; archaic mammals; dominance of "gymnosperms"; evolution of angiosperms; radiation of ammonites; "Mesozoic woodland revolution".
	Triassic		251	Continents begin to separate; marine diversity increases; "gymnosperms" begin to become dominant; diversification of "reptiles", including the first dinosaurs; first mammals.
PALEOZOIC	Permian		290	Continents aggregated in Pangea; glaciations; low sea levels; increase in "advanced" fish; different orders of insects; decline in amphibians; "reptiles", including mammal-like forms, diversified; major mass extinctions, especially of marine life, towards the end of the period.
	Carboniferous		354	Formation of Gondwana and the first northern continents; extensive forests of the first vascular plants , especially lycopsids, sphenopsids, ferns; first orders of winged insects;

			various angibia; first reptiles.
	Devonian	409	Diversification of bony fish; various trilobites; origin of ammonoids, amphibians,
			insects, ferns, seed plants; mass extinction at the end of the period.
	Silurian	439	Diversification of agnates; origin of fish mandibles (acanthodes, placoderms, osteocytes); the first terrestrial vascular plants, arthropods and insects.
	Ordovician	500	Divers ification echinoderms, other invertebrate phyla, vertebrates agnates; mass extinction at the end of the period.
	Cambrian	543	Marine animalsand diversify: first appearance of most animal phyla and many classes within a relatively short period of time; first agnate vertebrates; various algae.
PROT EROZ OIC		2500	First eukaryotes (1900-1700 m.y.a.); origin of the eukaryotic kingdom; fossil footprints of animals (1000 m.y.a.); multicellular animals from 640 m.y.a., including possible Cnidaria, Annelida, Arthropoda.
ARCH AEAN		3600	Origin of life in the distant past; first fossil evidence at 3500 m.y.a.; diversification of prokaryotes (bacteria); a photosynthesis generates oxygen, replacing the previous oxygen-poor atmosphere; evolution of aerobic respiration.

Source: Adapted from Jablonski and Bussler (1996).

According to the National Curriculum Parameters (PCN's), in order for students to understand the processes of diversification of species from an evolutionary perspective, they need to understand the historical-philosophical dimension of the production of science. It is essential that they have an understanding of evolutionary mechanisms, but for this they need a basic knowledge of some areas of knowledge, such as palaeontology, embryology, genetics and biochemistry, cell biology and others. In addition, the PCN's corroborate that these themes should be worked on in Basic Education, so that students can understand evolutionary theories, such as the Natural Selection proposed by Charles Darwin, the concepts of adaptation and evolutionary mechanisms in a temporal, geological and evolutionary dimension (Brasil, 1997).

Palaeontology is an area of knowledge of great relevance within the Natural Sciences, as it enables the study of events on a scale of thousands and thousands of years. Understanding geological, palaeontological and geographical events reveals the dynamics of planet Earth in different periods of history (Carvalho, 2010). Teachers play an important role in the classroom in terms of students' acquisition of new knowledge. Scientific information circulates rapidly when it comes to aspects related to palaeontology. It is being portrayed all the time in the media, the cinema, in reports and even on display in museums. This combination of information can often lead to misconceptions and misinformation (Brito, Pereira and Silva Filho, 2016). It is therefore up to the teacher to act as an instrument of transformation in the teaching-learning process, through effective methodologies that seek to meet the student's needs so that they can build meaningful learning (Izaguirry et aL, 2013).

Palaeontology in secondary education is covered within the content of evolution and is often seen in a fragmented and decontextualised way. Authors such as Moura and Barreto (2003), Schwanke and Silva (2004), Mello and Torello-de-Mello (2005), Sarkis and Longhini (2005), corroborate that this problem may be associated with the inappropriate use of textbooks and/or the teacher's lack of preparation for the subject. In addition, the scientific language contained in textbooks can be far removed from the reality experienced by students in their daily lives, which can cause serious damage to the process of learning. It is important for teachers to be clear and objective when they are teaching content in the classroom, as there is a need to introduce teaching methodologies that arouse students' interest in learning scientific knowledge and that bring real meaning to their education (Carvalho, 2010).

It is therefore necessary for students to have a broad and systematic understanding of the relationship that palaeontology fosters between the study of fossils and the understanding of evolutionary themes. Fossil records provide evidence that the biological evolution of species took place on Earth. For this reason, the aim of this study was to analyse the perceptions of students at a public school in São João do Patos/MA, Brazil regarding the study of fossils for understanding evolutionary changes in biological populations.

Methodology

Sample characterisation

The research was carried out in the city of São João dos Patos/MA, a Brazilian municipality located in the state of Maranhão (Figure 1). It is located 540 kilometres from the capital São Luís, in the Alto Itapecuru plateau micro-region, bordering the municipalities of Passagem Franca, Nova Iorque, Pastos Bons, Barão de Grajaú, Paraibano and Sucupira do Riachão with the Parnaíba River (Ibge, 2010). [a]The research was carried out with 33 (thirty-three) high school students at the Centro de Ensino Dr. Paulo Ramos (CEDPR), located at Rua Floriano Peixoto, n° 20, CEP 65665-000.

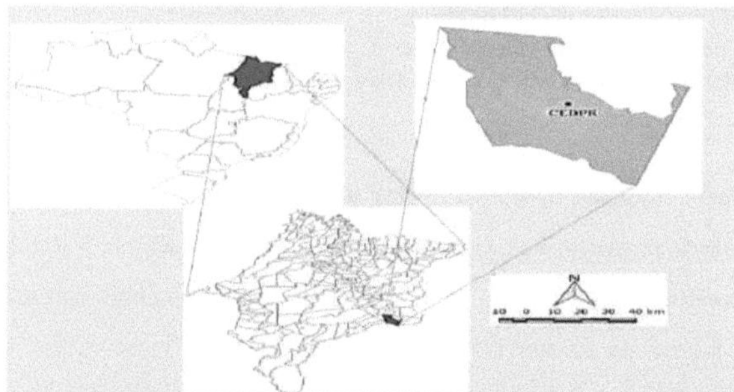

Figura 1. - Map of Brazil with an enlargement of the state of Maranhão and the city of São João dos Patos identifying the location of the Centro de Ensino Dr. Paulo Ramos (CEDPR), MA, Brazil.

During the methodological process, a qualitative and quantitative descriptive approach was adopted. According to Severino (2016), it involves a set of methodologies, instruments, investigations and references to the epistemological foundations of the study. The aim is to gain a close understanding of the *locus* in which the participants are inserted, in order to understand their ideologies, questions and point of view, so that all intrinsic and extrinsic aspects are presented. In addition, a bibliographical study was carried out on the subject in books and academic papers published in the main scientific journals and databases, which were made available on the main *sites* and databases: *Bioline International, Directorv of Ooen Access Joumals, Google Scholar, Scientific Electronic Library Online (SciELo), ScienceDirect, PubMed, Web of Science* and others.

Data Collection and Analysis

Data was collected using descriptive and systematic analyses. To do this, a form containing open and closed questions was used. This instrument sought to understand how the students perceived the dynamics of the Earth, with regard to the study of rocks for understanding evolutionary mechanisms. It also sought to identify the possible contributions of fossil records to the understanding of the subjects of Geography (Geology) and Biology (Evolution).

The data was tabulated and analysed in two separate stages. The aim was to identify the most relevant aspects for the research, first analysing both the qualitative and quantitative data using different measurement instruments. The qualitative analysis was carried out using the Content Analysis proposed by Bardin (2011), which comprises a set of techniques and systematic procedures that aim to identify, by describing the content of a message, aspects that can be inferred about the relative knowledge of the production/reception of the message. This content analysis was grouped into categories in order to identify the qualitative aspects of the research:

I- Satisfactory is when the answer shows the capacity for scientific argumentation, synthesis and logic using concepts discussed in the literature;

II- Partially satisfactory in some way, they present knowledge on the subject under investigation, but there is no construction of systematised scientific thought;

III-satisfactory refers to explanations that have no scientific basis, often linked to common sense, religious beliefs, sometimes without any meaning, because they are far removed from scientific knowledge.

Due to ethical issues and the categorisation proposed by Bardin (2011), the names of the students have been suppressed throughout the text in order to preserve their mentions and ideological positions. For this reason, their answers will be presented followed by the letter (E= student), preceded by a number such as: E1, E2, E3 and so on throughout this study, taking into account the order of presentation.

The quantitative data was analysed using the *Statistical Package for the Social Sciences (SPSS)* 23.0. This type of analysis is widely used in the social sciences, as it seeks to characterise the data found in the research through a quantitative assessment (Meirelles, 2014). This tabulation identified simple descriptive analysis, arithmetic mean, frequency, standard deviation and correlation analysis with the identification of *Spearman*'s and *Pearson*'s significance. The graphs and tables were built in the *LibreOffice®* Free programme using the *Cale* spreadsheet to make it easier to understand the results presented throughout the text.

RESULTS AND DISCUSSIONS

Thirty-three (33) students from the Centro de Ensino Dr. Paulo Ramos (CEDPR), aged between 15 and 18, took part in this study, the majority of whom were 17 years old. With regard to the distribution of students by gender, 58% were female and 42% male.

Palaeontology content is being discussed in basic education, mainly through certain areas of knowledge, such as biology, geography and history. In view of this theoretical/scientific apparatus and the palaeontological findings left by ancestors over the years, the students' understanding of fossil records was questioned, and the answers were categorised according to Bardin (2011). It was found that most of the students had a satisfactory knowledge of the subject under study, and that a small proportion were closer to the conceptions proposed in the textbooks (Table 1).

Table 1. - Students' perceptions of the fossil record at the Centro de Ensino Dr. Paulo Ramos (CEDPR), São João dos Patos/MA, Brazil, 2017.

Categories	*CEDPR students' perceptions*
Satisfactory	"They are the remains of beings that died a long time ago and through chemical processes, these remains have been preserved for many years under the Earth." (E1) "It's evidence left by ancestors that helps us identify the species that lived in the place where the fossil was found." (E2) "I believe that fossils are very important, because through them we can understand our past and the future, and also know what that being was like, how it lived, what happened in its time, how it died, etc." (E3) "These are remains of bones, pieces of pottery or even pieces of cloth that prove the existence of some organism and have been studied over the years. They are requirements that prove that certain beings lived on Earth thousands of years ago." (E4)
Partially satisfactory	"They are records, bones of living beings that lived millions of

Unsatisfactory	years ago." (E5) "The remains of an animal or plant that existed millions of years ago." (E6) "They are the remains of animals that lived millions of years ago." (E7) "I understand that a fossil is something we find of other living beings" (E8) Fossils are records that allow us to understand why an animal dies, and from there the fossil begins to evolve." (E9) "They are species that didn't undergo natural selection and as selection and the passage of time moved on, many didn't manage to stay on Earth for a while." (E10) "I don't know" (E11)

Palaeontological study through fossil records has enabled researchers to understand certain evolutionary processes with regard to species, fauna and flora. These records are not restricted to bones left by ancestors in rock structures. Fossils can be the remains of dead teeth or shells, dinosaur footprints, petrified trees and impressions made by soft bodies, such as jellyfish (Paulino, 2005). Because of this, fossils are of great importance for understanding evolutionary aspects. They are considered traces of animals or plants that enable the ability to characterise structures preserved in rocks, as well as identification in other media, such as ice, amber and asphalt (Cassab, 2010).

The aim was to find out, from the point of view of the students at the CEDPR school, whether the biology teacher worked on geological principles in his lessons and whether he made it possible to understand palaeontological and evolutionary aspects. It was found that 82 per cent of the students felt that the teacher explained where fossil records can be found. However, 3% said that in their teaching practice the teacher only explains the concept of fossil, without systematising the aspects that precede this content. For 6% of the students, the teacher does not explain any geological aspects in order to understand palaeontological and evolutionary issues. In addition, it was clear that some did not want to answer and/or left the question blank (Figure 2).

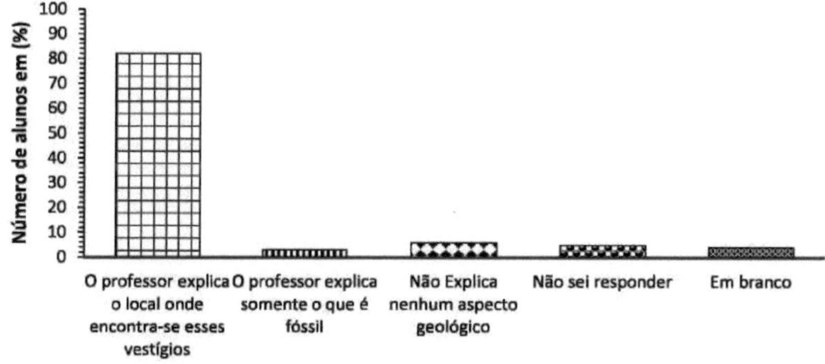

Figura 2. - Perceptions of students at the Centro de Ensino Dr. Paulo Ramos (CEDPR) regarding whether the Biology teacher explains everything from the geological aspects where fossils are found to evolutionary principles,
São João dos Patos/MA, Brazil, 2017.

In their teaching practice, biology teachers should work with integrating themes such as the origin of life, biological evolution of species, identification and diversity of living beings. In addition to the ability to articulate the content discussed in the classroom and understand the origin and evolution of life on a holistic scale (Brasil, 2014). For this reason, in order for students to be able to develop all this scientific apparatus, it is necessary for the biology teacher to work systematically, especially on geological subjects, so that they can understand the dynamics of fossils, in terms of the type of rock they are found in, the changes undergone by the relief, the incidence of weathering agents and the dynamics of the Earth. In this way, they will relate geological and palaeontological content to form their knowledge of evolutionary issues and thus develop their learning (Novais et al., 2015).The biological evolution of species is the central theme within Biology and there is a lot of evidence to prove this process (Reece et al., 2015). We therefore sought to identify the relevance of the study of fossils for understanding evolutionary processes from the students' perspective. Of these, 91 per cent reported that fossil records make it possible to understand the real existence of species in the environment in which they were collected and, through these observations, to identify successive evolutionary changes over the years. However, 6% said that these records are not related to evolution and 3% were unable to answer (Figure 3).

Figura 3. - Perceptions of the students of the Centro de Ensino Dr. Paulo Ramos (CEDPR) on the relevance of the study of fossils for understanding the biological evolution of species, São João dos Patos/MA, Brazil, 2017.

After the publication of *the* book *"The Origin of Species"*, the concept that beings are mutable was unveiled by Charles Darwin, but due to religious issues there is a certain resistance that lasts for years (Pegoraro et al., 2016). Palaeontology as a science seeks to elucidate evolutionary and temporal aspects in order to broaden understanding of the origin and evolution of organisms on Earth, as well as to identify through fossil records the changes that living and non-living beings have undergone over the geological eras (Mendes, 1986; Cassab, 2010).

Charles Darwin's theory of evolution was built on evidence observed in nature. For this reason, "he spent the next two decades gathering all the facts he could about botany, zoology, embryology and the fossil record." (Griffiths, 2013, p. 628). One of the great legacies left by Darwin for posterity was the explanation of evolutionary mechanisms. These were correctly incorporated into hereditary mechanisms, as they portrayed the transformations that species underwent over time (Griffths et aL, 2016).Biological evolution is characterised by a set of changes at the genotypic level that affect individuals in a given population. For this reason, we sought to find out how the students at the aforementioned school perceived evolution and the factors that affect these organisms. 67% of those surveyed considered evolution to be a slow and gradual process. On the other hand, 12 per cent saw evolution as a fast process. In addition, 3% considered these transformations to be a process that only happens in nature and 9% were unable to answer and left the question blank (Figure 4).

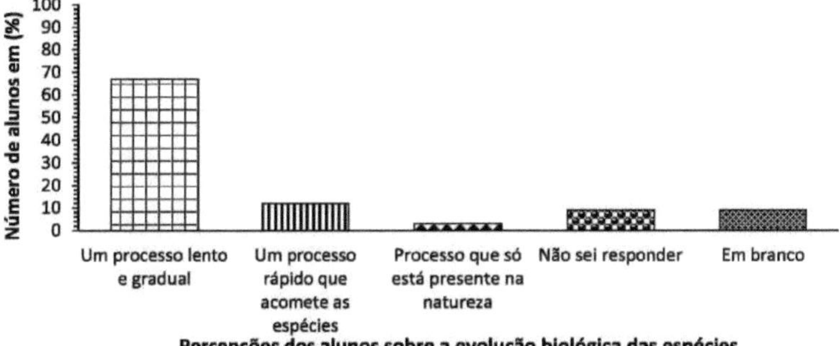

Figura 4. - Students' perceptions of the evolutionary process at the Centro de Ensino Dr. Paulo Ramos (CEDPR), São João dos Patos/MA, Brazil, 2017.

There are factors that tend to increase the variability of species, such as mutation and permutation, and those that act on genetic variability, such as genetic drift, migration and natural selection. For this reason, we sought to find out the students' knowledge of the evolutionary factors that can act on populations. In general, the students had some difficulty associating evolutionary factors with changes at the phenotypic level (Table 2).

Table 2. - Conceptions of students at the Centro de Ensino Dr. Paulo Ramos (CEDPR) about the evolutionary factors that act on a biological population, São João dos Patos/MA, Brazil, 2017.

Categories	*CEDPR students' perceptions*
Satisfactory	"Natural selection, mutation in genes that contributes positively to their survival." (E12) "Environmental factors, changes brought about by environment and mutation." (E13) "0 being slowly evolves and leaves the next more evolved generations (...)." (E14)
Partially satisfactory	"The natural environment and evolutionary processes." (E15) "The habitat, the climate, and the need to survive, to find food. These factors, such as the climate and the constantly changing environment, are the most likely." (E16) "It's the genetic characteristics that come about through natural selection." (E17)
Unsatisfactory	"It depends on the need to evolve, the more you evolve the more you do something" (E18) "Technology, knowledge, the genetic changes in each body, and this is evolution." (E19) "They are fossils or remains left in rock material." (E20) "Care and good nutrition." (E21)

"I don't know." (E22)

Biological evolution happens randomly and therefore changes can be positive or negative depending on the environment in which they are inserted (Ridley, 2006). In the book *"Each Case, One Case... Pure Chance: the biological processes of living beings"*, author Fabio de Melo Sene discusses some evolutionary factors such as: mutation, which is the only primary source of the appearance of genetic material, as it is considered a random event that changes the nucleotide sequence of the genetic material; the hitchhiking effect, which is provided by *linked* genes; recombination in meiosis, which is a reorganisation of the variables contained in the chromosomes or even by the formation of inversion, translocation, fusion/centric breakage and duplication or deletion of a piece of chromosome. Genetic drift is the change in allele frequency, as it is a micro-evolutionary mechanism that can modify this frequency over time (Sene, 2016).

In addition to the evolutionary factors mentioned above, migration also acts to change the genetic make-up of a population. For example, around the 19th century in a certain region of Europe, it was shown that gene flow increased due to the use of the bicycle, as it made it easier for people, especially men, to move around the region (Ridley, 2006). Another evolutionary factor is Natural Selection, which can change the allelic frequency of individuals, resulting in adaptation, an improvement in the average capacity of members of the population, which can provide the ability to survive and reproduce in the environment (Futuyma, 2009).With regard to the study of fossils, we sought to find out the students' perceptions of the area that has fossil records as its object of study. Around 91% reported that the purpose of Palaeontology is to study fossil records, 6% consider Ecology to have this duty and 3% reported that medicine is responsible for this study (Figure 5).

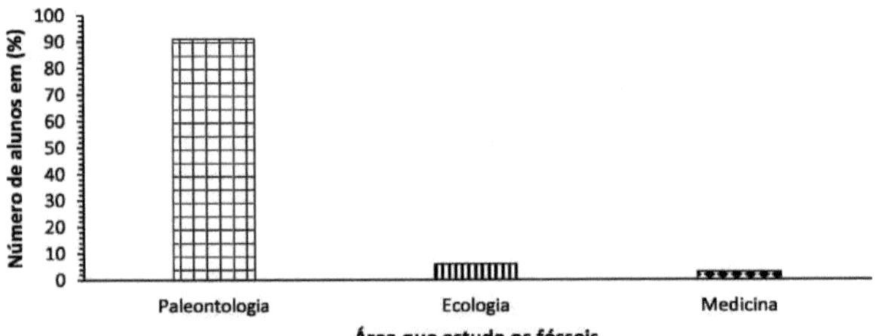

Figura 5. - Perceptions of students from the Dr Paulo Ramos Teaching Centre (CEDPR) about the ways in which they heard about fossil records, São João dos Patos/MA, Brazil, 2017.

According to the conceptions presented by the students, there was a statistical correlation between the question which sought to find out how the students perceived the biological evolution of species and the area of the study of fossils. Spearman's correlation showed a moderate negative correlation ($P=-0.531$; $p<0.001$).

The fossil record serves as evidence of the existence of species that have lived over the years and through these palaeontological finds it is possible to identify the impact of evolutionary factors on species. In this way, the students presented the contributions of fossil records, the emergence of species, the importance of rocks for the conservation of fossils and more:

> *"Fossil records are extremely important because they help us find a more rational answer about certain processes."* (E23)
>
> *"It's through fossils that we can understand what it was like before and understand the evolution of species."* (E24)
>
> *"Rocks help us to know the age of fossils, because these records are deposited in them and preserved for thousands of years. " (E25)*
>
> *"Through fossil records we can find out how certain species arose. "* (E26)
>
> *"Rocks are of great importance both for dating the age of fossils and for studying biological aspects of species."* (E27)

Fossils are portrayed in many mass media. The students were therefore asked in which media they had heard about this subject. Television (films, documentaries, reports, etc.) were mentioned by 43% of the students. 27% mentioned school, 21% books, 3% the internet and 6% said they had learnt about the subject from various other means of information (Figure 6).

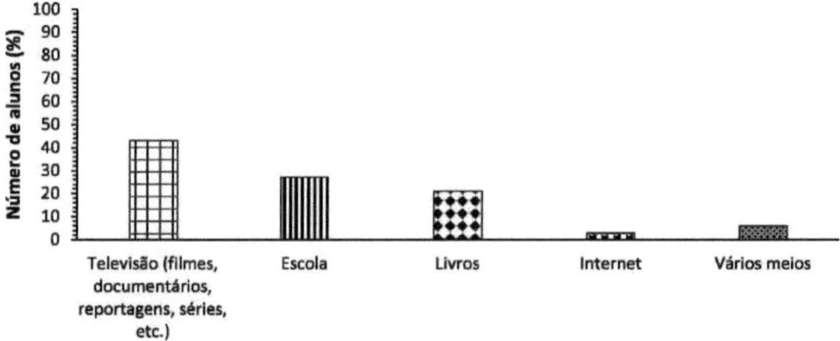

Figura 6. - Perceptions of students from the Dr Paulo Ramos Teaching Centre (CEDPR) about the ways in which they heard about fossil records, São João dos Patos/MA, Brazil, 2017.

School should be the main place where scientific knowledge is taught. The mass media serve as an auxiliary tool in this pedagogical process, but the school, through the contents of the common curriculum, should provide students with the ability to learn about these concepts (Carvalho, 2010). When it comes to palaeontology in basic education, there is a fragmentation of the theoretical/scientific apparatus in the high school curriculum, but the importance of museums and the media as a means of supporting students in learning more about palaeontological knowledge has been discussed (Schwanke and Silva, 2010).

Palaeontology is a subject that can be instigating for students if it is taught in a way that correlates theory and practice. We therefore sought to identify the importance of this area of knowledge from the perspective of the high school students at the school we surveyed. They presented the importance of palaeontology associated with the study of geology as areas that work together to understand macro and micro aspects of evolution:

"It's important to know which types of rocks contribute to the conservation of fossils." (E28)

"When we study rocks we study a bit about palaeontology." (E29)

"In most discoveries, fossils are found in rocks." (E30) "It's important to know the age of the fossil and the mechanisms of preservation over the years." (E31)

"It's important for a better understanding that the fossil found in the rock remains there for years." (E32)

"To study the rocks first and then the fossils, in other words, the types of rock that help preserve it." (E33)

For years, palaeontological knowledge was restricted to museums, research centres and academia, and in many cases the school community was distanced from this invaluable scientific knowledge (Schwanke and Silva, 2010). However, museums, the media and tourism play an important role in publicising and promoting prehistoric knowledge, as they show the history of the living beings that inhabited the Earth for thousands of years. Therefore, this scientific apparatus is of fundamental importance for students, including high school students, to see palaeontology from an educational perspective, not limited to a simple fossil, and it is necessary to include it in the Brazilian school curriculum (Carvalho, 2010).Regarding the didactic-pedagogical resources that biology teachers use in their classes, specifically when explaining the content of evolution. 64% of the students presented the textbook as the only teaching resource used in class, 3% said that the teacher uses *slides* in some classes, 27% couldn't answer and 6% left the question blank (Figure 7).

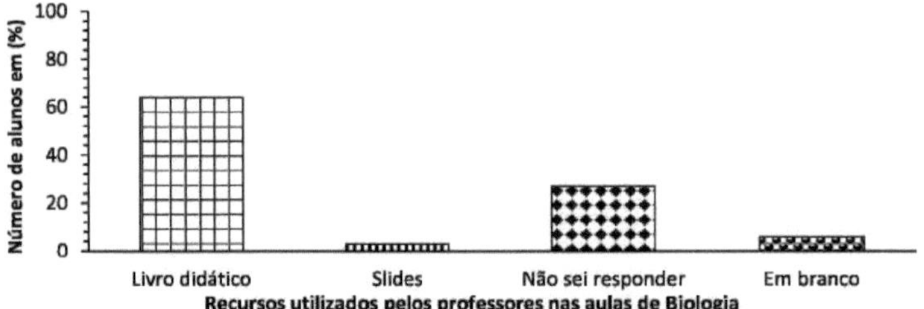

Figura 7. - Perceptions of the students of the Centro de Ensino Dr. Paulo Ramos (CEDPR) about the resources the teacher uses in Biology classes, São João dos Patos/MA, Brazil, 2017.

Palaeontological knowledge is almost always compulsory in Geology and Biology courses, but in many cases the subject is taught in a basic way, without going into greater depth on the subject. For this reason, the university is the ideal place to create innovative methodologies and develop instructional materials, breaking away from simplistic conceptions where teachers often only use lectures and textbooks as the only teaching tool (Carvalho, 2010).

CONCLUSIONS

It was observed that the contents of the area of palaeontology in basic education are dealt with in an insipid way, lacking theoretical and scientific depth, as far as the pedagogical practice of many teachers is concerned. Palaeontological knowledge is dealt with superficially in biology textbooks, specifically within the subject of biological evolution, which implies the need to look for various resources that promote the construction of knowledge on the subject. It was found that the majority of students at the Dr. Paulo Ramos Teaching Centre have a satisfactory conception of fossil records and their relationship with Charles Darwin's theory of evolution.

The students already understand the importance of geological study with regard to the dynamics of the Earth, specifically the study of rock types in order to understand fossil records. Even though teachers don't use technological resources in their lessons (a reality in most Brazilian schools), they still manage to achieve their students' learning objectives. The CEDPR students were able to perform well on the subject, especially in terms of geology, palaeontology and evolution.

It is therefore necessary for teachers to promote interdisciplinary work in their training so that students can develop work related to Geology, Palaeontology and Evolution. This knowledge is of great importance for students' cognitive development, so that they can demystify information linked to common sense and get closer to the scientific knowledge worked on by the teacher.

References

Alonço, M., and Boelter, R. A. (2016). Palaeontology in high school biology textbooks. *Revista da SBEnBio,^(9),* 7671-7682.

Araújo Júnior, H. I., and Porpino, K. O. (2010). Analysis of the approach to the theme of palaeontology in biology textbooks. *Anuário do Instituto de Geociências- UFRJ,* 33(1), p.63-73.

Bardin, L. (2011). *Content analysis.* São Paulo: Edições 70.

Brazil (2014). Ministry of Education. *PNLD textbook guide:* Biology. Brasília: Ministry of Education, Secretariat of Education.

Brazil (2016). Ministry of Health. National Health Council. *Resolution no. 510, of 7 April 2016. Official Gazette of the Federative Republic of Brazil,* Brasília, DF, 44-46.

Brazil (1997). Secretariat for Basic Education. *Parâmetros curriculares nacionais:* arte Secretaria de Educação Fundamental. Brasília: MEC/SEF.

Brito, L. A. R., Pereira, E. M. O., and Silva Filho, W. F. (2016). Public school students' knowledge of geology and palaeontology: challenges for scientific and professional popularisation. *Revista Encontros Universitários da UFC,* 1(1), 3687-3687.

Carvalho, I. S. (2010). *Palaeontology:* concepts and methods. 3. ed. Rio de Janeiro: Interciência.

Cassab, R. C. T. (2010). *Palaeontology.* Rio de Janeiro: Interciência.

Dobzhansky, T. (1973). Nothing in Biology Makes Sense except in the Light of Evolution. The *American Biology Teacher,* 35(3), 125-129.

Futuyma, D. J. (2009). *Evolutionary Biology.* 3. ed. Ribeirão Preto: FUNPEC.

Griffiths, A. J. F., Wessler, S. R., Carroll, S. B., and Doebley, J. (2013). *Introduction to Genetics.* 10. ed. São Paulo: Guanabara Koogan.

Griffiths, A. J. F., Wessler, S. R., Carroll, S. B., and Doebley, J. (2016). *Introduction to Genetics.* 11. ed. São Paulo: Guanabara Koogan.

Grotzinger, J., and Jordan, T. (2013). *Understanding the Earth.* 6. ed. Porto Alegre: Bookman.

Ibge (2010). Brazilian Institute of Geography and Statistics. *2010 Census,* 2010. Retrieved from: http://cod.iboe.Qov.br/15E2.

Izaguirry, B. B. D., Ziemann, D. R., Muller, R. T., Dockhorn, J., Pivotto, O. L., Costa, F. M. Alves, B. S.,

Ilha, A. L. R., Stefenon, V. M., and Silva, S. D. (2013). Palaeontology at school: a playful and pedagogical proposal in schools in the municipality of São Gabriel, RS. *Cadernos da Pedagogia,* 7(13), 2-16.

Joblonski, S., and Bussler, C. (1996). *Workflow Management Modelling Concepts, Architecture and Implementation.* London, UK: International Thompson Computer Press.

Meirelles, M. (2014). The use of SPSS *(Statistical Package for the Social Sciences)* in Political Science: a brief introduction. *Pelotas,* 14(1), 65- 91.

Mello, L. H. C., and Torello-De-Mello, F. (2005). Palaeo(e)geography: new challenges for teaching. In: BRAZILIAN CONGRESS OF PALEONTOLOGY, 19, Aracajú, 2005. CD, *Abstracts...* Aracajú, UFS.

Mendes, J. C. (1986). *Basic palaeontology.* São Paulo: USP.

Moura, G. J. B., and Barreto, M. A. f. (2003). Analysis of the degree of approach to the subject of palaeontology in high school biology textbooks. *Palaeontologia em Destaque,* 1(44), 1-6.

Muscente, A. D., Schiffbauer, J. D., Jesse Broce, J., Laflamme, M., 0'donnell, K., Boag, T. H., Meyer, M., Hawkins, A. D., Huntley, J. H., Mcnamara, M., Mackenzie, L. A., Stanley Jr, G. D., Hinman, N. W., Hofmann, M. H., and Xiao, S. (2017). Exceptionally preserved fossil assemblages through geologic time and space. *Gondwana Research,* 1(48), 164-188.

Novais, T., Martello, A. R., Oleques, L. C., Leal, L. A., and Rosa, A. A. S. (2015). The inclusion of palaeontology in primary education in different regions of Brazil. *Terrae Didática,* v.1, n.11, p.33-41,2015.

Paulino, W. R. (2005). *Biology.* 3. ed. São Paulo: Editora Ática, 2005.

Pegoraro, A., Soares, L. G., Rizzon, M. Z., Molin, E. D., Fernandes, F. M., Lovato, L. B., and Cunha, G. F. (2016). The importance of teaching evolution for critical and scientific thinking. *Interdisciplinary Journal of Applied Science,* 2(2), 1-6.

Reece, J. B., Urry, L. A., Cain, M. L., Wasserman, S. A., Minorsky, P. V., and Jackson, R. B. (2015). *Campbell's Biology.* 10. ed. Porto Alegre: Artmed.

Ridey, M. (2006). *Evolution.* 3. ed. Porto Alegre: Artmed, 2006.

Sarkis, M. F. R., and Longhini, M. D. (2005). A reflection on geoscience content in science and geography textbooks. In: BRAZILIAN CONGRESS OF PALEONTOLOGY, 19, Aracajú, 2005. *Abstracts CD,* Aracajú, UFS.

Schwanke, C., and Silva, M. A. J. (2010). Education and Palaeontology. In: Carvalho, I. S. *Palaeontologia:* conceitos e métodos. 3. ed. Rio de Janeiro: Interciência.

Sene, F. M. (2016). *Each case a case...pure chance:* the processes of biological evolution of living beings. 2. ed. Ribeirão Preto: Brazilian Society of Genetics.

Severino, A. J. (2016). *Metodologia do trabalho científico.* 24. ed. São Paulo: Cortez.

Zucon, M. H. (2011). *Introduction to Palaeontology.* São Cristóvão: Federal University of Sergipe- CESAD.

MIX
Papier aus verantwortungsvollen Quellen
Paper from responsible sources
FSC® C105338

Printed by Books on Demand GmbH, Norderstedt / Germany